MAIMAN'S INVENTION OF THE LASER

How Science Fiction Became Reality

Never Before Told Stories from the Early Days of the Laser

and the legacy of the company that Maiman founded, Korad

by Rod Waters

ISBN: 13:9781492842309

Library of Congress Control Number (LCCN): 2013921233

This book is based both on my personal experiences and interviews of over forty former employees of Korad. The book is as accurate as our memories permit. I view this work as a contribution to the early history of the laser. Therefore, it is as factual as we could make it.

This is a self published book.

Cover design by Dianna Little
Edited by Sharon Honeycutt, Anthony Simpson and Thomas Hauck
Images by Valentina Strvalia
Interviews transcribed by Lisa MacDonald
All above are available on elance.com

For more information:
Website: www.maimanbook.com

Preface

By Kathleen Maiman

As many of us already know, laser technology that began with Ted Maiman's first device has vastly changed our world, becoming as ubiquitous as the wheel. Ted's invention reflected the man himself—curious, practical, creative, and persevering. While others were pursuing complex gaseous designs for the laser, Ted's approach displayed his innate desire for simplicity. And simplicity was the recipe for his genius as inventor as well as his success as a full human being. He had an ability to combine the theoretical with the applied, and an ability to clear away all that was not essential to the problem at hand. In whatever he pursued, Ted was relaxed but focused and persistent.

As a person he was respectful of those around him, and he embodied a sense of dignity and integrity that not all in the scientific community can claim. In his work he wanted to help create a better world, as the endless medical, industrial, scientific, and consumer applications of the laser have indeed brought to mankind.

The laser, as one of the most important 20th Century inventions, ranks high because of its pervasive uses in such places as hospitals, homes, factories, farms, construction, war, the Internet, laboratories, and even in automobiles.

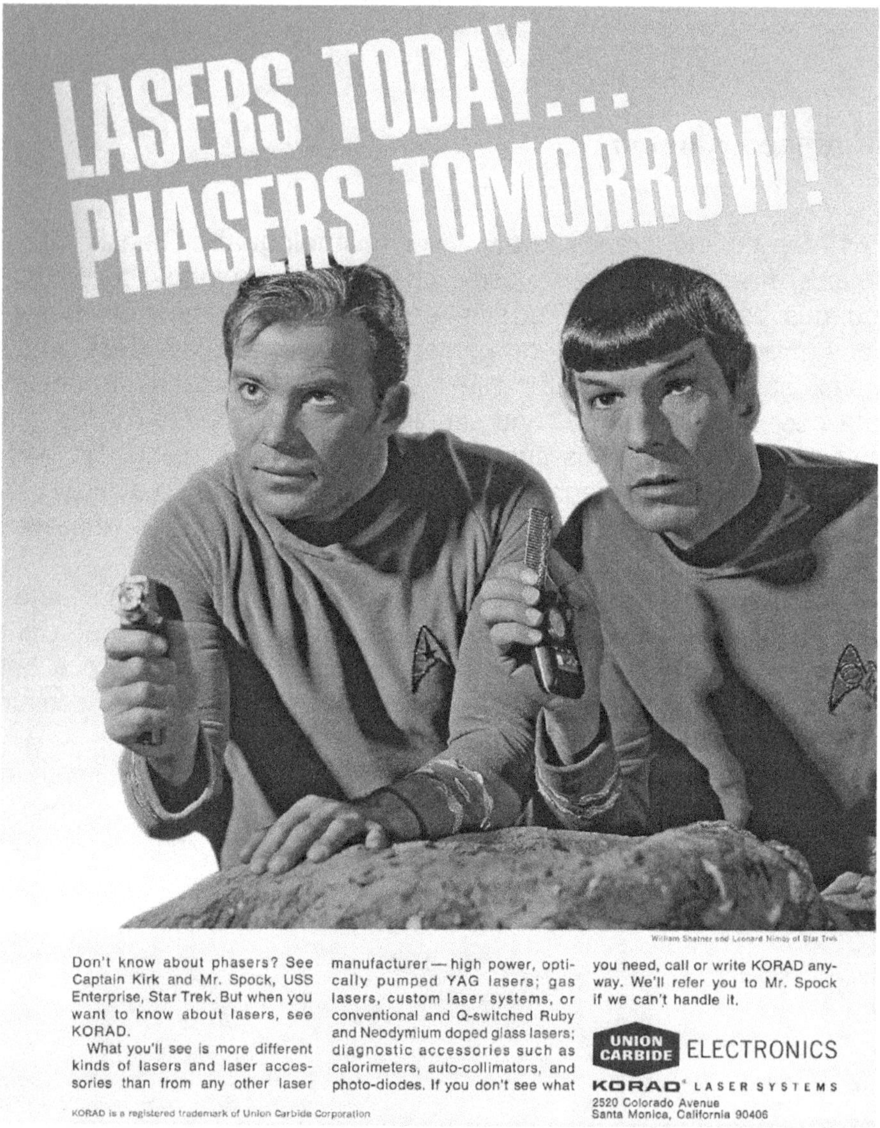

LASERS TODAY... PHASERS TOMORROW!

William Shatner and Leonard Nimoy of Star Trek

Don't know about phasers? See Captain Kirk and Mr. Spock, USS Enterprise, Star Trek. But when you want to know about lasers, see KORAD.

What you'll see is more different kinds of lasers and laser accessories than from any other laser manufacturer — high power, optically pumped YAG lasers; gas lasers, custom laser systems, or conventional and Q-switched Ruby and Neodymium doped glass lasers; diagnostic accessories such as calorimeters, auto-collimators, and photo-diodes. If you don't see what you need, call or write KORAD anyway. We'll refer you to Mr. Spock if we can't handle it.

KORAD® LASER SYSTEMS
2520 Colorado Avenue
Santa Monica, California 90406

KORAD is a registered trademark of Union Carbide Corporation

Korad's Most Famous Ad

Ran in many issues of Laser Focus magazine 1968-69

A full-color hi-rez enlargement of this colorful ad is available at

www.maimanbook.com

Acknowledgements

I am deeply grateful to Dr. Bill Buchman for the idea of writing this book and for reminding me to do it over a number of years. He never gave up. And to those forty or so Koraders throughout the United States who made this book possible by consenting to be interviewed and voting on cover designs—I am exceedingly grateful. Without them this would not have been the unique book that it is.

Then there are the many friends and relatives who read and commented on this book. First and foremost is my companion of the last twenty years, Bernice Ross, fellow author, with whom I attended about a dozen classes on how to write, most notably UCLA's Writing School in nearby Westwood. My closest family members provided encouragement over the years: daughters Danae Carney and Julie Egleseder and son John Waters. Cousin Betty Waters edited both this book and my previous one on the genealogy of the Waters family. Thanks for extra special help go to Hal Walker, Carl Schulthess, Cliff Cordy, Tim McGrath, Jeff Hecht, and technical editors Bill Rundle, Acle Hicks, and Don Roberts.

Special gratitude for writing and editing goes to Anthony Simpson, without whom the book could not have existed. I thank Fran McConnell for help on the book title and back cover. I wish to thank Dr. Walter Perrie for encouragement and editing. Talented eleven year old Joshua Galst pointed me towards simplicity and his father, David, provided valuable advice. My good friends, Jim Kessey, Jerry Shane and Marv Sachse were supportive and helpful.

Finally, I'd like to thank Joan Bromberg for writing a very thorough book on the early history of the laser.[1] It gave me insight into laser science history. Another useful resource was the Niels Bohr Library of the University of Maryland where interviews of the early laser pioneers are archived.

[1] *The Laser in America, 1950-1970*, Joan Lisa Bromberg, (The MIT Press, 1991)

In Lasers, KORAD® covers the spectrum

-from one end to the other, with more different kinds of lasers and laser accessories than from any other laset manufacturer. At various points along the KORAD spectrum you'll find high-power, optically pumped Yag lasers; gas lasers; custom laser systems; conventional and Q-switched Ruby and Neodymium glass lasers; and laser accessories such as calorimeters, auto-collimators, and photo-diodes. For a complete spectrum analysis, call or write KORAD today.

A full-color hi-rez enlargement of this colorful ad is available at
www.maimanbook.com

Contents

Introduction

Rodney (Rod) Waters, 1969

Here I am, seventy-seven years old and living in an age where sophisticated technology is everywhere. I marvel constantly at how what was once considered the stuff of Buck Rogers or *Star Wars* is now as ubiquitous as the toothbrush.

I worked for the inventor of the first laser, Dr. Theodore (Ted) H. Maiman during the pivotal years of the mid-sixties and early seventies at the company he founded, Korad Lasers, in Santa Monica, California. Maiman's ruby laser was the first device to successfully generate light in a coherent beam.

Although I was five years too late to witness the invention of his historic laser, I was fortunate to work for him from 1965 to 1968, where I established my life's work in lasers. I witnessed remarkable feats accomplished by those who worked under him at Korad. This book is about their incredible adventures. Those were the days of international acclaim and wonder.

Today, when I am passing through the checkout at computer stores, I often see laser pointers capable of beaming light across a room or a parking lot into a tiny dot. How curious that a phenomenon so taken for granted today was the stuff of science fiction not long ago. All of us at Korad felt we were doing important work; we were benefitting humanity, ourselves, and having an exhilarating time.

Back then we never thought lasers would be in virtually every home. We considered them to be scientific instruments with a potential only for industrial work like welding and drilling. Maiman himself said he never expected them to be in everybody's home. Now they inhabit every computer as hard drive read/write heads and every video (DVD) player. And of course, we could not have imagined the Internet compressing the entire world into a village. Lasers carry the world's communications—including telephone, email, and TV—through fiber optic cables over land and beneath oceans. Of course we knew of light's enormous information-carrying capacity, but we never anticipated a networked world. One only hopes that lasers will someday be used to ignite thermonuclear micro-explosions in fusion reactors, thus helping to solve the energy requirements of the human race.

Even though we hadn't the least notion of these and other spectacular applications for this technology, we Koraders were determined to be the first to demonstrate the laser as a practical reality. That is what this book is about: a fevered race to beat our various competitors to the punch; to carve out for ourselves a bit of glory and perhaps acclaim; to get up every day and head to work in a fascinating area of endeavor, working with an eclectic group of personalities, and ending up in places and situations we could never have anticipated.

Theodore (Ted) H. Maiman

Dr. Maiman finally won the race to invent the first laser,[2] besting his more establishment opposition, a development which rankled his competition immensely and resulted in some rather creative accounts of his struggles for recognition. The details of his battle will be reviewed herein as it is a component of my overall message, but it will not be discussed exhaustively as Maiman himself covered this quite well in his autobiography *The Laser Odyssey*,[3] currently sold from Maiman's website, LaserInventor.com, and also on Amazon.com.

Further exploration of Maiman's battle for recognition with Dr. Charles Townes of Bell Laboratories was most conspicuously provided by Jeff Hecht in *Beam: The Race to Make the Laser*, published in 2005 by Oxford University Press. It is my goal here to flesh out those tightly focused accounts with a more general story of the laser industry of that time; the people involved, with all their quirks and eccentricities; the adventures we stumbled into; and various other dynamics heretofore unexplored. In fact, this book is currently the world's only record of the exploits of the people of Korad Lasers, Maiman's *second* creation. I thought it important to get the history of Korad down for the record, for we Koraders are all now in our seventies and eighties and the clock is ticking down.

---

[2] Maiman received a patent on his ruby laser; therefore he was a "true" inventor.

[3] Maiman, Theodore H. *The Laser Odyssey*. Laser Press, 2000.

Not Out to Pasture Yet

This is not, however, merely the decades-eroded recollections of an old man, musing from an easy chair. Indeed, it's far from it, for although I am three-quarters of a century in age, I work full-time doing what I love (working with lasers), plus I enjoy excellent health and can even occasionally be found on the highest ski slopes. The research for this work took the form of interviews with those involved—the scientists, salesmen and technicians[4] I worked with during that period, over forty of them altogether. Reconnecting with them at their homes at various locations throughout the United States was like going back in time to the days of my youth at Korad in Santa Monica, California. Here were my fellows from the heart of California's bustling scientific community of the sixties and seventies. Here again were my friends, so many years removed, opening once again to me like favorite old books, still containing the same treasured stories page after page.

What This Book Is About—In a Nutshell

The stories of many Koraders are told here, such as Hal Walker's wild ride down from Lick Observatory on Mt. Hamilton when rednecks tried to run him off the road to a sure death. He succeeded in beaming a Korad laser off of a reflector placed on the Moon by Astronaut Buzz Aldrin of Apollo 11 fame. Another rousing story tells of Carl Schulthess' involvement in the Vietnam War's first use of a "smart" laser at night to "paint" targets on the Ho Chi Minh Trail for destruction by Air Force jets. Schulthess himself narrowly escaped his own destruction by a North Vietnamese suicide squad. The use of these weapons in winning the Gulf Wars of the early 1990s is discussed.

After leaving Korad, Cliff Cordy got his doctorate in engineering and tells amusing stories of Engineers Gone Wild—techies at play both during and after work. In spite of what people may think of us engineers (nerds), Korad's favorite restaurant for entertaining customers featured nude dancers.

Anybody who buys a diamond needs to know about lasers used since the 1960s to enhance these gemstones so that they sparkle more ~~and thus cost more. However~~, their value is diminished because drilling

[4] Interviewed also were four of Korad's company presidents: Bill Thurber, Don Sims, Bob Schlesinger, and David Collins. Unfortunately, two others, Ted Maiman and Clay Zerby, are deceased, and Schlesinger died in 2010.

diamonds with a laser to remove imperfections weakens the stones and reduces resale value. The holes cannot be seen and sellers are not required to disclose. Buyer beware!

Industrial lasers like welders, drillers and cutters play an important role in the history of the laser. Stories of their early days are in the book.

We wrap the book up with a bang, describing the scandals from Ronald Reagan's presidency, such as the October Surprise and the infamous Inslaw Affair, both of which were directly related to Korad.

Dr. Maiman's struggles for recognition for having created the laser more than fifty years ago are central to the book. The story of his conflicts with his arch rival, Dr. Charles Townes, is told. After a half century, we can say he won his battle. Although he died in 2007, his widow, Kathleen, is still fighting for him.

Ted and Kathleen Maiman

What possessed me to hop about the country on jet after jet, in search of these interviews? It certainly wasn't the desire to unburden myself of some disposable income or acquaint myself with the airlines' various movie offerings. Certainly I could have gathered the details through email exchanges, phone conversations, or some other more affordable means, but that was only part of the point. I missed those faces, those voices, those quirks, and nothing short of a physical reconnection over a coffee or snacks would do. The result of those interviews was an imposing stack of recorded material subsequently transcribed to several thousand pages and used to compile this volume. From the people directly involved in the race to create the laser, you will hear a compelling story. I hope you enjoy reading about it as much as I enjoyed being a part of it.

Readers uninterested in details of the invention of the laser[5] might turn to Chap. 4, Spying and the First Billion Watt Laser.

---

[5] Lasers are almost always uniform both in space and in time. This means uniform cross sections of the beam in space and uniform waves in time. Another word for "space" is "spatial" and another for "time" is "temporal." Soldiers marching in step at a parade can be said to have similar spatial coherence in their orderly formation and temporal coherence in the timely beat of their boots.

Chapter One: "DEATH RAY" Bragging Rights

There has been a major disagreement in the scientific community as to who invented the laser. This conflict arose when Maiman was working as an obscure, West Coast scientist at Hughes Research Laboratories in sunny Malibu, California. Some discussion of that conflict is in order before we discuss Korad specifically.

In order to determine who deserves the title of inventor of the laser, it is important to clarify what *inventor* means. Thomas Edison serves as a prime example. He was prolific, for in addition to the incandescent light and motion pictures, he held patents on a thousand other gadgets. So we see that inventors hold patents, but what exactly is a patent? Essentially it is a license granted by a national government to sue in a federal court. An individual holding a patent can sue another who infringes on his patent by stealing his idea and using it for financial gain. Lawsuits, however, are enormously expensive, and beyond the means of most small companies. So why go to the expense of getting a patent? Bragging rights are one good reason, but a better one is that a patent greatly increases the sales price of your company should you want to sell, especially if you want to sell to a large company. Needless to say, big companies have lots of money, so they can not only afford to defend their patents but also to drive out competition.

The situation is further complicated by today's system of information exchange—specifically the Internet. Now that people are able to purchase products from any number of locations around the globe, how can a single patent in just one country protect an inventor? Unfortunately, it can't. If an inventor wished to ensure complete protection, he would have to get a patent in *every country*, about 160 of them. But as a practical matter, an inventor generally seeks patents only in the largest countries. With that brief explanation of the nature of a patent, let's discuss the invention of what was, at its debut, melodramatically referred to as "The Death Ray."

The Media Has a Field Day

"LA Man Invents Death Ray!" was the July 1960 headline in the *Los Angeles Times* heralding a scientific breakthrough of the highest order. As dramatic as that sounds, the actual term was LASER, an acronym for Light Amplification by Stimulated Emission of Radiation. Since that time lasers have been employed to perform a variety of tasks, including eye surgery, tattoo removal, computer read/write devices, and most importantly, making the World Wide Web a reality. Oh yes, and lasers are used in industrial tools like welders, cutters, markers and drillers. What a career we had, contributing to making some of this possible!

The Laser Started with Uncle Albert

The laser actually began with Albert Einstein, who in 1917 predicted stimulated emission as a new process in the interaction of radiation with matter. (Remember that "stimulated emission" is the "SE" in LASER.) It is the process by which light can be amplified to enormous levels, far brighter than the sun. This incredible energy can be harnessed in a sufficiently controlled manner so as to allow its use in supermarket scanners without the shopper leaving the store blinded for life. Lasers come in all shapes and sizes, like ants compared with elephants.

Gordon Gould

Art Schawlow

Charles Townes

While searching the Internet, we find that a fellow named Ted Maiman holds a patent on the ruby laser, but not *the* laser. That honor belongs to Gordon Gould. But wait a minute! Arthur Schawlow's obituary claims he is the co-inventor of the laser with Charles Townes.

All four of these men lay claim to inventing the laser; what's going on here? To help us answer that question we will need to go back a few years to 1953 and discuss the development *of the first* Microwave Amplification of Stimulated Emission by Radiation, or MASER. This breakthrough, made by Professor Townes of Columbia University, proved that Einstein was correct; stimulated emission of radiation really could be produced in a device. Before Townes made the first emitter of stimulated emission, an ammonia maser, some of his colleagues had entertained doubts as to whether stimulated emission was even possible because prior to that only spontaneous emission[6] had been observed.

Maiman, in his book *Odyssey,* takes pains to point out that Townes' maser put out only one ten-billionth of one watt. Maiman was perhaps too generous because he failed to point out that his original ruby laser produced one hundred million times more power. He did, however, pithily observe: "In the end, although the maser provided a very interesting bit of physics exploration[7] for several years, it was no more than an interlude, if not a distraction, on the way to the laser. It was plainly a backward move from the work of Fabrikant."[8]

In 1939[9] Dr. Valentine A. Fabrikant, a Soviet Russian, was the first to come up with the *idea* of a laser as a device to produce coherent light. Even though his concept was explained in his doctoral thesis and in a Russian patent, it never resulted in a working model and received little recognition. In spite of that, he is generally credited in the literature with his idea of associating Einstein's theory about stimulated radiation with a device that actually emitted it.

---

[6] Spontaneous emission fills the universe; it is the light from our Sun, the stars, candles, and all other sources except lasers, masers, and some interstellar molecular gas clouds.

[7] Because of the limited practical uses for the maser, it became known as Means for Acquiring Sums for Expensive Research, according to Dr. Jim Boyden, Korad engineering manager.

[8] Maiman, *The Laser Odyssey*, 59.

[9] V.A. Fabrikant, Doctoral *Dissertation,* FIAN P N Lebedev Physical Institute, Academy of Sciences, USSR (1939)

Seven years after Townes' invention of the maser came the first demonstration of the laser as a coherent light amplifier by Ted Maiman. His discovery was just in time as scientists were once again concluding that stimulated emission of light might not be possible. Microwave,[10] yes. But light? Certainly not! Universities, including Columbia and Stanford, had been conducting research at a feverish pace in an effort to develop this technology. In addition, American industries, including Hughes Aircraft, IBM, AT&T, Raytheon, and others, were participating in this effort and expending huge heaps of cash. After all, our dreaded adversaries, the Russians, were wasting no time developing their own Death Ray. Matter of fact, two Soviet Russians shared the Nobel Prize[11] with Townes for the microwave laser (maser).

Ignorance and Arrogance in Academia

Dr. Maiman's paper announcing the first laser was turned down by *Physical Review Letters*, the premier physics journal in America. Maiman had titled his paper "Optical Maser Action in Ruby," so PRL's editor thought it was just another of the many papers on the maser. Maiman wrote a letter to the editor, Sam Goudsmit, who again refused to publish the paper. Because of this second turndown, in his memoir, Maiman accused PRL of "shenanigans," alleging a deliberate attempt to steal his discovery and attribute it to others such as Townes and Townes' brother-in-law, Dr.** Arthur L. Schawlow. He harped on them in his book and in life, calling them the "East Coast Establishment." He points out that PRL published all laser papers after denying his, especially the ones written by IBM and Bell Labs.[12]

Maiman therefore acted quickly and sent a short paper to the British Journal, Nature, which was published on the 6th August, 1960. This is one of the most important papers in the history of optical science.[13] It ignited a frenzy of inventions of lasers of many different types such as gas, liquid and more solid types.

---

[10] "Microwave" as in oven, radar, cell phones and FM radio.

[11] The Russians developed the theory while Townes got stimulated emission working. Townes' achievement was a world first, thus proving the concept was true - for which he got the Nobel Prize.

[12] Maiman, *The Laser Odyssey*, 112.

[13] *"Stimulated Optical Radiation in Ruby"*, Maiman, Nature, Aug 6, 1960, vol. 187,

My friend Dr. Walter Perrie, laser scientist at the University of Liverpool in England, weighed in on that question, telling me he was unconvinced that Maiman had been thus frustrated by Townes and Schawlow, saying, "I have always held them in high regard as the people whose work led to the laser." However, in his memoir, Maiman disclosed that Townes was on the review board of PRL and thus, Maiman felt, in a position to deny his publication. Besides, Schawlow had another working ruby laser cobbled together in only three weeks at Bell Labs and was positioned to grab the credit after the news of Maiman's invention of the ruby laser had reverberated around the world. Perrie did note, however, that based on Maiman's book, *Odyssey*, "It appears there were attempts to negate and minimize Maiman's achievements for which, I believe, he should have received the Nobel Prize."[14]

Despite the fact that he was denied the Nobel Prize, Maiman's accomplishments have been heartily recognized by major scientific awards. Further, he was made a member of the Royal College of Surgeons in Britain, the only non-surgeon to be so honored. Dr. Perrie concluded: "In the end, does it really matter? Maiman is famous for the invention as well as Schawlow and Townes. They all contributed in a human endeavor to push light amplification to the optical region. All were brilliant, and all of humanity has benefited as a consequence of their efforts to create a totally new industry.

(4736), 493-494.

[14] Personal communication.

Ted Maiman with his original tiny ruby laser[15]

Consensus with Qualifications

As a general rule today, few will categorically deny Maiman credit for inventing the world's first laser; it's just that it's often mentioned equivocally, as in: "Maiman developed the first *ruby* laser" or "the first *working* laser" or "the first *pulsed ruby* laser," as if there were some other laser that was invented before his. "Working" is the wording of those who would minimize his achievement in hopes of ensuring for themselves a place in history. In addition to those mentioned above, Maiman experienced fierce competition for the invention of the laser from Bell Telephone Laboratories of the mighty American Telephone & Telegraph (AT&T) Company's high citadel of "pure" science. When I was a teenager back in 1950, I regarded Bell Labs as number one in the world of pure science. "Pure," as explained to me back then by my eighth-grade science teacher, meant they did only basic research without regard to mere product development.

---

[15] Maiman's laser used a 25mm long synthetic ruby crystal from Union Carbide, produced a laser pulse of 10 kilowatts peak power (e.g., 5 joules in a 500 microsecond pulsewidth) in a beam of low divergence. Such light had never been seen before.

Bell Labs invented the transistor, giving rise to the microelectronics and Internet revolution of the late twentieth century. Bell then aspired to be honored for having invented the laser based on their 1958 publication of a paper on the MASER by Townes and Schawlow—technology they subsequently patented. In their patent documentation they claimed an optical maser was an imminent possibility ("optical" meant "light," the "L" in LASER). As a result, after this patent was awarded, Bell Labs temporarily received credit for the invention of the laser, but the patent was later struck down by a federal court, thus negating their claim to fame as inventor of the laser. However, Maiman did receive a patent on his ruby laser, which underscores this book's title.

Perhaps these modifiers of the historical record would have us refer to the Wright Brothers as merely building the first *working* airplane or Edison as developing the first *working* light bulb. Bell Labs did invent the first *nonworking* laser—or did they? We discussed Valentin Fabrikant, the Russian who received a patent for conceiving of a device to amplify light by stimulated emission in the 1930s. Of course, it did not then carry the name *laser*; it was Gordon Gould who is said to be the first to apply that label by replacing the "M" in the well-known term, maser, with an "L." But this was not exactly a stroke of genius or originality, as the maser had already been invented by Townes.

As noted above, the person who is *granted a patent* on something is the inventor. You don't need to make your invention *work*; you merely need to describe or teach *how* it would work. Gould and three genuine scientists—Townes, Schawlow, and Maiman—fought over the right to be called the inventor of the laser. Perhaps "fought" is too strong a word for refined scientists. Let us use *sparred* or say they participated in *petty infighting* instead. To be sure there were a number of lasers invented after Maiman's ruby—many of them. Just a few were the gas lasers, helium neon, and carbon dioxide and the solid types like Maiman's such as semiconductor, diode, and optical fiber. It has been commonly said by laser folks that almost anything will lase if you hit[16] it hard enough. I recall a claim that a certain Scotch whiskey vapor had been made to lase.

---

[16] The technical term for "hit" is pump.

All These Recipes, But Where's the Beef?

Just as many inventors had labored to build an incandescent light before Edison, there were many individuals who contributed to the development of laser technology. However, the six most prominent scientists responsible for the invention of the laser were Einstein, Alexander Prokhorov, Nicolay Basov, Townes, Schawlow, and Maiman.[17] But like Edison, who built the first working light bulb, Maiman invented the world's first working laser. Patent law aside, that was one hell of an accomplishment! Certainly it deserves more than an obscure cobwebbed mention in the history books. He proved to the world that coherent light resulting from stimulated emission really did exist. (There were doubts that it did!)

Maiman's invention triggered a frenzy of research on lasers in the scientific community. Gould, who was not a member of this community, nevertheless won the patent race with the help of powerful and well-paid attorneys. And if you equate invention strictly with *patent*, then Gould was the inventor. He and his lawyers collected approximately one hundred million dollars in royalties off the laser, many times the total amount received by the other patent holders *combined*.

One could say Gordon Gould invented the *first nonworking laser* as a result of a decision by a Federal court of law. Maiman called this a "travesty of justice" in his book.[18]

The second laser invented was by Sorokin and Stevenson at IBM in NY (November 1960). This laser came only six months after Maiman's in May of that year. However it was just a "knock off" invention as the IBMers simply replaced Maiman's ruby laser rod by a uranium one and used Maiman's flashlamp idea to energize it with Schawlow's mirrors to amplify it. Only a few months later Ali Javan's nifty gas laser arrived, the helium-neon, known as a HeNe (he-nee).[19] This was to become the heart of the ubiquitous supermarket scanners and laser pointers. As it was a continuous laser, not pulsed as Maiman's and Sorokin's were, it

[17] Gould is generally recognized as an engineer, not a scientist.

[18] Gould won the patent lawsuits after eighteen years of legal battles, a "travesty" in Maiman's opinion (*Odyssey*, 186).

[19] Maiman's original laser was about ten million times more powerful. Such is the nature of lasers, there are all sorts of them.

could also carry information. It was said back then one HeNe laser beam could carry all the television stations on Earth!

Genesis of Genius, Maiman the Boy Engineer

Maiman's career success was due primarily to his father, Abraham, who, as an electronics engineer and inventor, began training his son in electronics at an early age. Indeed, young Ted was so proficient at electronics that it helped finance his way through college. His father instilled in him an approach to problem solving and invention which enabled him to find novel solutions. Through sheer determination and a belief in himself, he was able to win in spite of many factors working against him. In *Odyssey* he cautioned that one should not always follow conventional wisdom and that no matter how talented the individual is, he or she can still get things quite wrong.

As another example, Dr. Walter Perrie cited the famous British scientist Lord Kelvin, for whom the Kelvin temperature scale was named. In spite of being a brilliant physicist, Kelvin was noted for getting many things wrong and, for example, did not believe in James Clerk Maxwell's electro-magnetic theory. Kelvin entered Glasgow University at age ten!. Becoming professor at age 22, he famously concluded that the Sun could be no more than one million years old in disagreement with geologists of the day. But he succeeded in developing the first and second laws of thermodynamics which say that perpetual motion machines are impossible. Or simply put, you can't get something for nothing.[20]

It is generally felt that Maiman's hands-on, get-it-done approach to life allowed him to win the race to invent the laser. With his father, his master's degree in electrical engineering, his doctorate in physics, and his experience at Hughes Research Lab building an ultra compact microwave laser—the maser—he was well equipped to be the tortoise amongst hares in this race. None of his rivals held a master's in engineering. This distinction, according to Korad scientist Bill Buchman, PhD, from Cal Tech, was the factor that enabled Maiman to win. In short, he was an engineer as well as a scientist.

---

[20] The three laws of thermodynamics are like Isaac Newton's three laws of motion, all are laws of the Universe.

The Establishment Cedes a Belated Acknowledgment

The world now marks the laser's birthday as May 16, 1960, the day on which Maiman first operated his laser, and 2010 marked its fiftieth anniversary. Throughout 2010 there were birthday observances at many universities worldwide as well as trade conventions, such as one I participated in at San Francisco, Photonics West. There were dozens of photos covering the exhibition's walls, boasting the accomplishments of the various laser pioneers, and many articles in the science magazines. Kathleen Maiman, Ted's widow, also delivered an excellent speech at a laser conference in Paris in conjunction with the observance, in which she very ably defended her late husband's place in history. Professor Townes gave a stately, somewhat more subdued presentation at the same function. But then Dr. Townes, a man now well into his nineties, gives a fascinating contrast.

Based on the overall festivities of the fiftieth anniversary of Maiman's achievement, it seems the fight over who invented the laser has finally resulted in his winning his struggle for recognition.[21] The laser is one of the most amazing inventions of all time, and as we can now appreciate, a lot of people deserve credit for having provided a significant service to humanity.

Both Thomas Edison and Ted Maiman not only invented, they also commercialized their contributions as businessmen. Both established companies that further developed their inventions. The company Maiman founded, Korad, produced and sold the world's most powerful and most reliable lasers during the decade of the 1960s. As an example, consider the Korad ruby laser at the McDonald Observatory in Texas, which fired countless billion-watt laser pulses at the moon for fifteen years from 1970 to 1985, obtaining scientifically valuable ranging data.

Oh, but I'm ahead of myself. For the moment, let's focus on the next phase of our story: Wow, this is one cool new toy. Now what can we do with it?

---

[21] Maiman's winning recognition is not completely recognized on the Internet where "noise" abounds. You can check this out by googling "laser inventors."

Chapter Two: Korad Lasers, The New Kid on the Block

After his triumph at Hughes, Ted Maiman's career took some sudden and precarious turns. The realization that such a device was now reality caught the attention of the scientific community worldwide and set off a mad scramble to refine and expand on Maiman's baby. He himself found this intensely competitive environment distasteful; the inherent political maneuverings that accompany large research projects were to him a significant distraction from the pure pursuit of science. Further, as he notes in Odyssey (158), his relationship with George Birnbaum, head of Hughes' research department, was anything but cordial. It seemed as though the same dynamic was at work at a couple other companies he had considered, and Maiman began to wonder how he was going to manage in this new environment.

The government's Advanced Research Projects Agency (ARPA) was responsible for developing new technologies for the United States military. As the research arm of the Pentagon, it had, about a year before Maiman's laser invention, given a million-dollar contract[22] to Technical Research Group in New York City to develop the first working laser. TRG was one of Maiman's biggest competitors and had only asked ARPA for $300,000. That bid, a princely sum in 1960 dollars, was tripled by Uncle Sam, much to the delight of all those earning their living at TRG.

[22] It was the very first laser contract and was called Project Defender. Valued in 2013 dollars, it would be about $9 million.

Such wanton extravagance was unusual, even for the United States government, which is normally not heralded throughout the world for its spendthrift ways. But they could picture beam weapons! (ID the target, guide a smart bomb to it, and *kaboom!*) Then there were the possibilities of super accurate laser radars and even Death Ray Buck Rogers-type beam weapons, known then as "Star Wars" and now as *electric lasers*. Why, for the Pentagon brass, that was simply *irresistible!* Now, with Maiman's breakthrough, Pentagon engineers had a delightful new toy to play with. Government contracts flowed in a green, flood-engorged torrent to Maiman's employer, Hughes Aircraft, which was close to being a 100 percent military supplier.

Doing the Amazing as a Warm-up to the Incredible

Fran McConnell had this to say about Maiman's Korad: "I heard guys say that Korad is a place to discuss new scientific possibilities with others who understand you. They got to think outside the box – that's what Maiman did! He gave them a place to dream and he was proof that it could work. He wasn't the most personable man, he was shy. But he always said hello if he walked by the desk of a lowly secretary. I never forgot that."

Maiman Cuts the Maser Down to Size

His tenure at Hughes was marked by breakthroughs, such as his jaw-dropping redesign of a massive ruby maser, basically a microwave laser. This maser was an immensely cumbersome behemoth; if it had legs, it could surely have stomped through the suburbs and stopped Godzilla in his tracks. He reduced the size of this lumbering beast by a factor of *a hundred* (a 99 percent reduction), from approximately four hundred pounds to eventually four. Maiman accomplished this with his characteristic out-of-the-box approach to the problem. He thus made Dr. Townes' maser invention more practical, allowing for its use in such applications as the atomic clock. But more importantly, the experience he gained with his ruby maser helped in the invention of the laser.

When Ya Gotta Go, Ya Gotta Go

An opportunity to escape the environment at Hughes presented itself in early 1961 when Maiman received an offer from Leonard Pincus, formerly of Hughes himself, who wondered if Maiman might be interested in joining the team at a new company in Santa Monica,

California. Founded with venture capital, the company was to be called Quantatron. Being a "big frog in a small pond" appealed to Maiman at that point, and he accepted their offer, becoming vice president of Quantatron's Applied Physics Laboratory in April.[23] Morale at Hughes during this period was at low ebb, which made it relatively easy for him to recruit other employees to accompany him to Quantatron. In total, eight Hughes employees made the jump.

In *Odyssey*, Maiman writes, "Quantatron was a subsidiary of Union Texas Natural Gas Company. Venture capital monies had funded Quantatron, but the assets were not well managed and the funds were rapidly dissipated. Early in 1962, Allied Chemical Company acquired Texas Gas."[24] And with it came Quantatron.

According to Marv Sachse, Korad's sales manager, Texas Gas had invested $1.2 million in Quantatron improvements. (This would be worth $9.2 million in 2013.)

In-House Baubles and Pilfered Platinum

One critical goal Maiman had during his tenure at Quantatron was the in-house manufacture of ruby and other crystals. He preferred to be in control of this vital aspect of laser production as it would be preferable to relying on an industry competitor for these critical components. In the Second World War, the United States government persuaded the Swiss to transfer to them the ruby crystal technology that made Swiss watches tick so reliably. Union Carbide's Linde Division was the sole beneficiary of this transfer, employing it after the war to produce commercial synthetic gems. As gorgeous as these may have been, they did not have satisfactory properties to adequately function in Maiman's newly created ruby laser. This issue was promptly addressed by Ricardo (Rick) Pastor, a brilliant physical chemist and close personal friend who had accompanied Maiman to Quantatron from Hughes. They had met in the lobby while applying for positions at Hughes Research Labs (HRL). Later on Dr. Pastor was instrumental in the invention of Maiman's laser. They used to go together to the parking lot to inspect the latest rubies in the trunk of the Union Carbide Linde salesman's car. Pastor even helped Maiman choose the color of his ruby (pik)[25] which turned out to be the key to successful lasing.

---

[23] Maiman, *The Laser Odyssey*, 158.

[24] Ibid. 160

Ricardo (Rick) Pastor, 1977

Pastor's innovative techniques and painstakingly careful methodology resulted in the production of ruby laser rods that were, at least for the moment, superior to those produced by Linde. But Pastor's dominance didn't last long. Within a year, Dr. Ralph Wuerker was buying rubies from Switzerland by way of a Providence, Rhode Island, company, Adolph Mueller.

Wuerker was the designer and builder of the world's first best-selling laser, which came to be known as the Korad K-1. The Korad brand was to confer pride of ownership to the industry and university scientists who purchased it for their experiments. In 1965, the price of the K-1 basic laser was $9,250.[26] In 2013 this would be worth over $70,000.

---

[25] Personal communication (email) with Dr. Pastor's son, Louie.

[26] Editorial, *Laser Focus* (Oct. 1, 1965): 17.

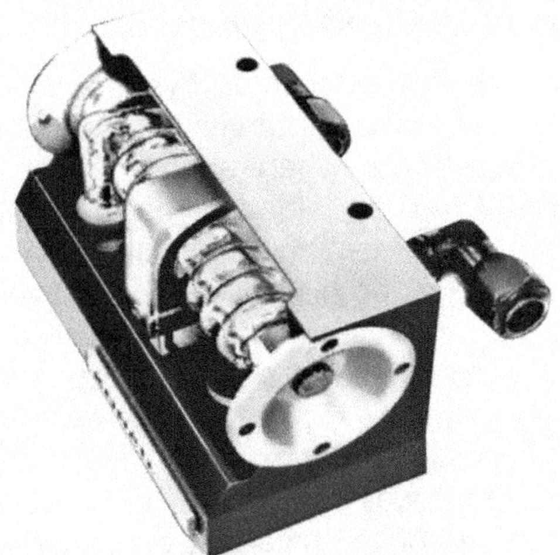

Cut away view of K-1 laser head showing Korad's unique spiral flashlamp and ruby rod peeking out the front. A simple laser, but expensive.[27]

Ralph Wuerker, father of K-1 laser, 1971

[27] Lasers tend to be expensive because physicists are!

Dr. Wuerker, the First Chief Engineer

An Occidental and Stanford educated physicist, Wuerker[28] was eighty-one years old at the time of our interview on April 2, 2010, at his home in Westlake Village, California, where a large fat cat strutted about as if it owned the place. A retired widower, he walked with the aid of a black cane. Wuerker knew Maiman personally, having studied with him at Stanford. He marveled at his colleague's maverick approach and stated so during our interview. Wuerker recalled being thoroughly impressed by Maiman's genius as a scientist, physicist and as an electronics engineer expert. "I think that maser that he made at Hughes was just brilliant," Wuerker said. "He did some really clever engineering there. And it was just typical of the guy. He was that way at Stanford. He was one of the smartest experimentalists in the whole Stanford department."

Not Your Father's Physicist

Another interesting insight offered by Wuerker was the difference between physicists of today and those who worked on the laser back in the fifties and sixties: "We were not stupid with our hands in those days. As a physicist in those days, in the fifties, by the time you got out of Stanford and Caltech and places like that you were expected to be pretty good with a lathe, with glassblowing, and with electronics. This was not the digital age; this was the analog age." Wuerker elaborated on this shift from a hands-on, tangible method to another, purely intellectual approach: "It's really bad that it's gone from the Socratic place to the Aristotelian approach. Think about it in your head and that's all you need to do. And if you can't reduce it to practice ...."

This is an interesting observation in light of the patent peculiarity discussed earlier, where knowing is everything and doing is practically nothing. Maiman was undeniably the first to reduce the laser to practice as an inventor, according to Wuerker who closed that particular musing with a recollection of going to cocktail parties at Maiman's home where he noted with curiosity the little ruby crystals his host had attached to one of the lampshades—rejects from his pursuit of the ruby maser.

---

[28] He passed away October 29, 2012, at age eighty-three. His obituary in the *Los Angeles Times* stated he held twenty-five patents and called him a "workbench" physicist.

Because of Maiman's independence and temperament, his life at Hughes was perhaps hampered by several difficulties, including friction between him and George Birnbaum as well as others in management. As Wuerker recalled, Maiman didn't like "people pushing him around too much."

Chasing the Federal Dollar

According to Don Roberts, one of Quantatron's early personnel hired as a junior physicist by Wuerker, the company stayed afloat in those early days by aggressively pursuing government contracts. Roberts was seventy-eight at the time of our interview, retired, and living in the San Diego area with his wife Lee. He said in our interview of February 21, 2010, "Dr. Bill Buchman was one of Korad's key people, and was very busy writing proposals that no one at Korad could understand because they were so sophisticated. They were theoretical contracts like you have in universities. Maiman told me he thought that even the people that paid for the studies didn't understand them."

Donald (Don) Roberts

William (Bill) Buchman

While Buchman was the primary mover behind these proposals, he also had help from others such as Bernie Soffer,[29] Ray Hoskins, and

[29] Bernie Soffer was the only one of these four without a doctoral degree, yet he was

the Pastor Brothers, Rick and Tony. Roberts explained, "Rick was in the same nucleus of brilliant people able to get so many government contracts to fund Quantatron." Roberts joined the company soon after Maiman took the helm. He was to be physicist Wuerker's assistant. "When I went in, they had everything set up, all the departments. Everybody was there. They even had a library that I used extensively, a very thorough library having a lot of very good research books." Roberts said, "Maiman ended up with a group of physicists who were all very good."

Along with brilliance often comes eccentricity, and Roberts recalled one of Wuerker's more endearing peculiarities: "Ralph had a million things going in his mind. His desk was piled high with papers. The interesting thing about Ralph was that if I went into his office and said, 'I need such and such a document,' he'd say, 'Just a minute.' Then he'd dive into that pile of papers, reach down underneath, and pull it out. He knew where everything was. He was very easy going, full of jokes, a funny guy. He looked something like the actor Stewart Granger."

A Yearning for Recognition

Roberts also reflected on his discussions with Maiman over the latter's frustration over having to fight for his due recognition as the laser's inventor. "He fought that battle over and over again. He used to come to my office and complain about Townes and how he, being on the review board for the *American Physical Review*, rejected Maiman's original paper, and so he had to go to the British *Nature* to get his paper published on the invention of the laser -- the first publication telling how a laser really worked."

This involves a critical argument of the period: can anyone claim credit for the invention of a device if they are only responsible for a *theoretical, written discussion of it*, without ever having actually reduced it to practice?

By Roberts' estimation, the work of Townes, as expressed in his famous *Physical Review* article,[30] failed to pass that test. "Townes'

---

one of Korad's top scientists. He contributed to the study of nonlinear optics, specifically frequency multiplication such as devices that could convert infrared lasers into green and ultraviolet ones.

[30] A.L. Schawlow and C.H. Townes, "Infrared and Optical Masers," *Phys. Rev.* 112,

paper did not. Maiman used to talk about how if you write a patent, you should be able to make the product from the knowledge of that patent." Roberts was adamant on the point, insisting, "From Schawlow and Townes' paper or patent, you could not reproduce a laser. The recipe was wrong."

Patent Regrets

Roberts said there was a lack of focus at Quantatron for the potential patents they could have claimed for the many innovations developed there. In support of that, he went on to say, "They didn't get the idea to patent things until I was there probably five years, then they came to me and said, 'Don, we've got to get patents. What kind of patents can you think of?' I was really surprised because we had done a lot of very original work. We designed and made all kinds of lasers using flash lamps and rods with silver coatings on the ends called Brasher's silver. We coated both ends and scratched a little hole in one end for the laser beam to come out."

Lamenting this particular missed opportunity, Roberts continued, "Coatings like silver on laser optics were originally a big problem, they wore off quickly ruining the laser. When they made them of multilayer magnesium fluoride and cerium oxide, all of a sudden the coatings were very hard and you couldn't scratch them. That was one of the real advantages of Quantatron at that point, they had a really tough, long lasting surface on them. All laser coatings since then have been the same. . . It should have been patented."

During that period, at the time of hire virtually all engineers and scientists in every company were required to sign an agreement relinquishing to the organization all rights to inventions. In general, inventors were not rewarded with a lot of money; they received "glory," some received bonuses, some stock options—it was all negotiable. Roberts noted the contrast between Quantatron's failure to pursue patents with that of his subsequent employer National Cash Register (NCR). "The method of having the laser cavity flooded with water, which is what we used in all our laser heads, was our idea and could have been patented. The whole seal system on the flash lamp could have been patented. I just didn't think about doing it. I didn't

---

1940 (1958). This famous paper inspired lots of scientists to join the race to invent the laser, including, of course, Maiman.

think it was that clever, but it could have been. I've had seventeen patents since I left Korad. When I was at NCR, they had a patent office right in the building."

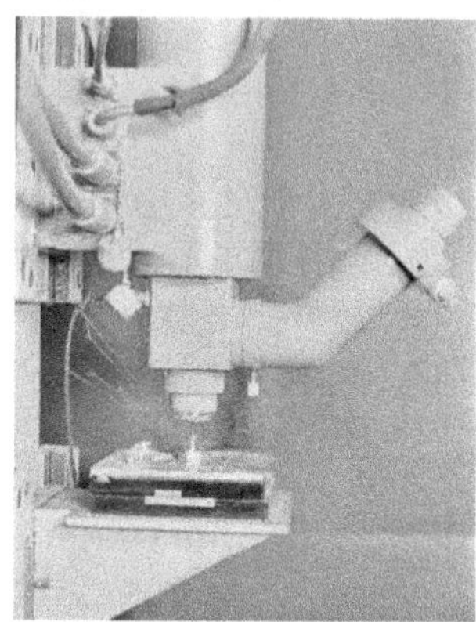

Early laser machine, 1963

The above photo shows one of the world's first laser machines. It is dated (by typewriter) at the year of transition from Quantatron to Korad. The photo originates from Maiman's papers in the possession of his widow, Kathleen Maiman, and is reproduced here with her permission.

Quantatron Becomes Korad

"One morning as I came to work, there was a man waiting outside," said Don Roberts. "As I unlocked the door he blurted, 'It's about time somebody showed up! I'm from Allied Chemical. I wanted to check into buying this place. But now I wouldn't touch it with a ten-foot pole. Nobody comes to work early and does any work around here!'"

Despite this lukewarm assessment of Quantatron's work ethic, the operation was purchased by Allied Chemical in early 1962, by which time Maiman had built his workforce up to thirty-five people. But Allied was to prove unenthusiastic where lasers were concerned.

Quantatron was involved in two main areas, lasers and microwave switches, which were really two unconnected companies under one roof. However, Allied seemed to Maiman to be focused mainly on the microwave part. Wuerker told me Allied purchased Quantatron so fast they didn't even know about the laser operation. While Maiman and his staff were realizing some significant progress, especially with Wuerker's new Q-switched[31] Model K-1 prototype, Maiman found Allied's less than halfhearted support disappointing and began searching about for a more supportive parent company. Union Carbide Corporation emerged early as a very interested party, and Bob Charpie, assistant to the president of Union Carbide in charge of corporate R&D worldwide, was the initial point man in the huge corporation's negotiations with Maiman.

Robert (Bob) Charpie

Looking to Gobble a Minnow

Charpie had met Maiman a couple years earlier when Hughes' soon-to-be laser inventor had approached him about the acquisition of synthetic rubies, which Hughes did not manufacture themselves. While Charpie was to find Maiman's negotiating style much to his eventual distaste, he reported being impressed at this initial stage by Maiman's "intense, if not passionate, approach." Hughes was uncooperative and refused to provide him the necessary assistance in his quest to produce the first laser, so Maiman asked at their first meeting: could Charpie possibly provide him with some of Carbide's baubles?

Now well into his eighties at the time of our interview[32] and recalling those days from the comfort of the home he shares with his

---

[31] A Q-switch is a device located between the laser mirrors that produces extremely high-power laser pulses, millions of watts. Scientists of the time were intrigued by this new tool. For more on Q-switches, read Appendix I.

[32] Sadly, he passed away in October 2011, just twenty-five days after his beloved wife, Beth. His obituary noted he received two Bronze Stars as a World War II soldier.

wife Beth in Weston, Massachusetts, Dr. Charpie recalled Maiman's frustration with his earlier experience at Hughes, when he had endured his employer's repeated refusals to provide him with large synthetic rubies. Charpie, however, not only sat on an advisory board to the Pentagon, he was also a member of the National Science Foundation and appreciated the laser's tremendous potential for use in advanced military applications.

Carbide had been in the ruby-growing business since the 1940s when the Nazis shut off the Swiss supply of synthetic rubies, and their crystal growing group was designated by Uncle Sam as the organization responsible for employing this Swiss technology to further our war effort. Up to that point, these gems had been used only for watches and precision military instruments. Indeed, for years Carbide enjoyed a most profitable business marketing a good portion of these ruby and sapphire gems in the commercial jewelry trade, profiting with a gross margin well up the 90 percentile. To Maiman, Carbide was an obvious choice of suppliers in his quest for bigger rubies with which to produce more powerful lasers. And subsequently, Carbide was to supply Maiman with the rubies that produced the world's first laser beam.

Bob Charpie in 1944 as a 19-year-old soldier

Hunker Down Boys

When it came to his approach to deal making, Charpie was no teddy bear, having achieved his position with Union Carbide as a result of his unusually high intelligence, tenacity, and drive. Indeed, during World War II, he had served in the Army's Tenth Mountain Division, the Ski

Troops, most notably (and historically) involved in a raging attack in February 1945 on Hitler's forces that were hunkered down on Mt. Belvedere in northern Italy.[33] In the Battle of Belvedere 1,000 Americans lost their lives out of 12,000 attackers, one of the highest ratios in the war.

In 1960 Charpie was head of all Carbide's research worldwide and reported directly to Carbide's CEO, Birney Mason. Carbide employed 200,000 people at the time. So there was a lot of research taking place, mostly at their Oak Ridge, Tennessee, nuclear facility which alone had 70,000 people, a third of the company. Oak Ridge was the biggest piece of Carbide, a veritable city 25 miles west of Knoxville, mostly involved in uranium purification for Uncle Sam.

Charpie had come to Union Carbide as the former director of Oak Ridge's Nuclear Reactor Division. He had resigned from that position as the business of producing commercial nuclear reactors had become a dying pursuit. He opted to join Carbide due to an irresistible offer that included an office on the forty-third floor of corporate headquarters in New York City.[34] As a nuclear physicist recently in charge of one of Carbide's R&D departments at Oak Ridge, he was well equipped for his new position.

Communications Versus Military Lasers

After Maiman's invention of the high-power ruby laser in May 1960, it was a whole new ballgame in the scientific community where laser research was concerned. Maiman's number one competitor and rival was AT&T's Bell Labs in Murray Hill, New Jersey, with Dr. Charles Hard Townes, their chief maser and laser scientist, who was assisted in his work by his brother-in-law, Dr. Arthur Schawlow. Bell's primary focus was on continuous-emission, low-power communications lasers like milliwatt helium neon models and their notorious failure, the potassium vapor behemoth. This focus on communications did not excite the Pentagon at that time. They wanted high-power, range-finding military lasers capable of determining the distance to a target as well as guiding munitions to that target. And, lucky for the Pentagon, the first

---

[33] In his Memoirs Charpie tells a story of how he and a buddy delivered an Italian woman's baby on their way to the Battle of Belvedere.

[34] Carbide headquarters were at 270 Park Ave. in Manhattan, New York, occupying the entire skyscraper.

laser to be invented was at a major military supplier, Hughes Aircraft. Remember all those amazingly accurate laser guided "smart" bombs we watched bombarding Baghdad on the evening news during the opening of Desert Storm in the Gulf War of 1991.[35] That sophisticated technology experienced its first infant-like movements during this period as various institutions employed their expert teams in a savagely competitive effort to realize hugely lucrative applications, both commercial and military, from Maiman's invention.

Quantatron was off and running under Maiman's leadership with three separate avenues being traversed: the refinement of his ruby laser under the direction of Wuerker, exploration into semi-conductor lasers under Dr. Fred Burns, and the CO2 gas laser approach in the hands of Dr. Ray Hoskins.

Wuerker reflected in our interview on one aspect of the government funding that Quantatron and others were competing for. Wuerker noted that both he, during his pre-Korad work at Ramo-Wooldrige (TRW), and Maiman, during his tenure at Hughes, were affected by a federal law passed in the 1920s, the Independent Research and Development Program (IR&D). As Wuerker noted, this federal mandate required aerospace companies to allocate five percent of their government contract funds to research and development. This requirement had a major effect on telephone giant AT&T, which for decades enjoyed a near monopoly. Basic research at Bell Labs in Murray Hill, New Jersey, was federally mandated, and the telephone rates were artificially pumped to very high levels. Businesses, especially, were subject to exorbitant rates, while private residences were kept at an artificially low rate; after all, businesses don't vote, people do. Wuerker noted, "It was the government that told Bell Telephone System to fund the research lab. That's the way you fund research in a pluralistic society. So after World War II and the resultant expansion of related industry, they passed the same rule on the aerospace industries."

---

[35] The loss of innocent lives in war is regrettable; fortunately it is somewhat preventable with targeted smart laser bombs and missiles—and also GPS and video-guided missiles. Hitting the target precisely then requires a smaller amount of explosives, thus less collateral damage.

Of Big Bucks and Big Brains

Bell Telephone Labs, having been thus funded lavishly by inflated rates and at government directive, as well as staffed by an impressive collection of the best minds in science, was well positioned to win the race to build the first laser. It came as quite a surprise and an even greater embarrassment when they were beaten to the punch by a relatively unknown West Coast scientist named Ted Maiman. Wuerker observes, "Another thing that I think made them mad is that there's always, in my opinion, been a tension between western physics and eastern physics. The East Coast thinks they're the smart ones and we're a bunch of cowboys out here."

It was in this harried research atmosphere that Maiman and his contemporaries found themselves during the decade of the fifties and beyond. And with only twenty-five employees and operating as a venture-capital gamble, Quantatron was but a tiny minnow in the sea. Wuerker offered a rather unflattering assessment of Quantatron as "a joke" that was on shaky ground upon his arrival—obviously on the verge of failure. But Maiman, for his part, was determined to forge ahead and be as productive as possible, and this required rubies, which brought him back in contact with Union Carbide's Bob Charpie.

In his attempt to persuade Charpie to approve the production of larger rubies, Maiman cited applications he was sure would spark Carbide's interest, including laser welding, which offered a good potential for innovation at Carbide's Linde Division where many types of welders were built and sold. At that point, the big boss Mason of the massive corporation of Union Carbide had not yet come to fully appreciate the sea swell of advancements that laser technology would soon provide and had to be led by the hand to embrace lasers, one of the more significant scientific developments of the twentieth century. And Charpie was just the man to lead him.

So Who Is This Maiman?

The first time Ted Maiman approached Bob Charpie was in the spring of 1960, when he was looking for tiny rubies for his laser invention at Hughes. The second time came in mid-1962 on behalf of Quantatron; this time he suggested to Charpie that perhaps he and Carbide would be interested in purchasing a really small company with lots of potential. From his consulting work at the Pentagon in Washington, DC, Charpie was well aware of the US government's intense interest in military lasers. And as a prominent scientist, businessman, and deal maker at Carbide's Manhattan headquarters, he was well aware of the commercial uses as well.

Charpie found his negotiations with Maiman to be less than smooth, for Maiman had proved to be inflexible where negotiating was concerned—and for good reason; he was well aware of the laser's potential and was likely not willing to part with Quantatron for less than a fair price. Charpie regarded Maiman as a poor team player; he was a loner, a scientist at heart who had no particular love or enthusiasm for the business end of running a company.

How would Quantatron size up as an addition to the mighty Union Carbide, a multibillion-dollar enterprise employing a few hundred thousand people, Charpie wondered. Certainly this would be the smallest by far ... but still, it was a promising new company with the famous Maiman leading the research on a cutting-edge technology. As time passed the idea seemed to make more and more sense to him, even though in 1962 there were not yet any commercial laser applications. But if Carbide were to be in the business of producing the rubies used in laser production, didn't it make sense for the company to expand directly into the production of lasers themselves? When he recalled the laser contract recently awarded to Technical Research Group in New York at three times over the amount TRG requested, Dr. Charpie began seeing dollar signs pulsing like a laser beam across his consciousness. If this stubborn Maiman fellow would just play ball!

How the First Laser Got Invented

Charpie wrote a book about his exploits and had it privately printed for his children and grandchildren. He gave me a copy in which he relates the details of Maiman's visit in early 1960 to his impressive office in Carbide's corporate skyscraper in Manhattan overlooking Madison Avenue:

"A young man from Santa Monica, California by the name of Maiman visited me in my New York office. He was working for Hughes Aircraft Company. I did not know who he was, but I was very impressed by his intense, if not passionate, approach. 'I need a big piece of solid ruby,' he told me. 'How big can you make it?'[36] We began to play around with the idea. We were able to grow a synthetic ruby of significant size. Later, I pick up the Sunday *New York Times* and read that Theodore Maiman of Hughes Aircraft had just made the world's first laser. It was made from a large synthetic ruby."

A second set of negotiations between Maiman and Charpie occurred two years later in 1962. In his unpublished autobiography, Charpie relates how he had decided to investigate getting into the sapphire bulletproof window business when Allied Chemical got in trouble with the US government for polluting the skies over their plant in Tonawanda, New York. This led Allied to divest itself of troublesome and low-profit businesses, including their sapphire operation. Dr. Charpie discovered that Allied also owned Quantatron and wanted to rid themselves of that pesky Maiman and his lasers. Quantatron was involved in the production of microwave switches as well as lasers, and Allied decided the laser business had to go; they just wanted the switches. Charpie jumped on the opportunity.

For Sale: Laser Business, Cheap

Charpie was glad he had refused Maiman's price and threw the negotiations over to Ward Moore, who eventually became his assistant in the new electronics division.[37] Moore finalized the deal, doubtless on excellent terms as Allied was a motivated seller. My opinion is that Carbide got a "steal" in the deal. My meager evidence is that none of my interviewees, who should have known the sales price, knew this closely guarded Carbide secret, and so it was most probably a steal. Also, Carbide was quite generous in giving the "Maiman Group" twenty percent ownership for free. Those "in the know" were Ed Young and Drs. Bill Buchman and Bob Charpie. I interviewed all three. Buchman was advisor, and he and Maiman frequently lunched together. Charpie was head of Carbide's R&D; without Charpie's approval, deep pockets, and generosity, Korad probably would not have been created and would not have flourished. Finally, Ed Young was Maiman's trusted business and financial advisor, a close friend, and a lifetime Carbide

---

[36] Quoted by permission of the Charpie Family.

[37] In my second interview of Dr. Charpie a few months after the first, he reiterated that dealing with Maiman was "basically a pain in the butt!" He went on to say Maiman's only interests in life were himself and getting a Nobel Prize.

"man." Any one or all of these men should have known how much Carbide paid Allied for Quantatron.

Charpie's decision to pursue the purchase of Quantatron from Allied Chemical was not made in haste. As mentioned, it would be the smallest "by miles" of the other operations he oversaw. In mid-1962, his research of the laser business had uncovered absolutely no commercial laser uses. Yet Carbide certainly could benefit by combining their growing of ruby and sapphire crystals (mostly gemstones) with an equally ambitious program to produce powerful lasers; applications were a matter of time and research. Carbide by then had the best rubies. Uncle Sam was willing to pay huge sums for laser weapons and had already done so with TRG, so the case seemed clear. Carbide purchased Quantatron from Allied Chemical in October of 1962, after which Maiman was named president of the new corporation. Maiman picked the name Korad, an acronym for COherent RADiation, with the C hardened to a K. All that was physically added at the time of Korad's founding was the name over the door at Quantatron's 25,000-square-foot facility at 2520 Colorado Boulevard, Santa Monica, California.

But on the inside, the change was much greater. Allied took out the microwave switch business, which freed up about half the building for laser work.

Chapter Three: A Lousy Bedside Manner?

Maiman, the reader will recall, was renowned more for his brilliance as an innovative scientist than for his skills as a smooth negotiator and team player. He experienced friction with Carbide's management as he had with his superiors at Hughes. His widow, Kathleen Maiman, touched on this aspect of her late husband's personality in an email with me on August 31, 2010:

"Carbide looked over his shoulder watching and would get angry when Ted did such things as tear out the time machines from the machine shop. But he reminded the corporate heads he could. Ted was not doing this from ego but from his belief that ALL his Korad group were of an elite breed. The creative talents needed to build new companies to develop such things as lasers don't get nurtured, but are easily destroyed by the rigid culture of large corporations like Carbide and Hughes. Too many empty suits and investment bankers insisting on quick profits and unreasonable return of investment run large enterprises."

Milt Laikin, now owner of Laikin Optics in Marina Del Rey, California, was one of Maiman's crew of brilliant engineers, supplying practical optics expertise he had taught to the troops during the Korean conflict. In those days all the big artillery guns were sighted optically, and it was Laikin's job to instruct others in the repair of optical equipment from small binoculars to big range finders.

His experience under Maiman resulted in decidedly mixed emotions. Reflecting on those days in an interview at the Marina City Club in Marina Del Rey, on February 6, 2010, Laikin discussed his admiration for Maiman as a scientist, saying, "I very much admired him; I just never *liked* him." While Maiman may not have been universally loved, he was a very brilliant engineer and physicist, who chose to regularly visit all Korad's engineers and discuss their projects with them. Laikin again: "It was almost as if when he spoke with me, he understood the optics of the Kerr cell and Pockels cell Q-switches[38] that I was developing. Then if he would talk to somebody else, he understood that. So he understood a huge amount of technology and was fluent enough to discuss all the details with all the engineers, which in that respect I thought was very remarkable."

Laikin's difficulties included frustration with Maiman's management style, which he felt kept those involved in production off balance. "We never could get out of him how many units he wanted. In other words, if the company was going to manufacture one Q-switch, that's one thing. If the company wanted to make a product out of it to sell 100 of them, that was an entirely different story. Never could get out of him what volume or what his marketing plan was." Korad was working on a number of government projects, and Laikin commented on his impression of Maiman's approach: everything built to order, with inadequate focus on spare parts to be stocked for the lasers produced. "They would have a government contract to make whatever the hell it was, and that was it; there was no tomorrow. But then you have to advertise that you have such-and-such a thing. You have to have a marketing strategy anyway." However, Laikin's point of view is that of an engineer. A business person would view it as keeping inventory cost low.

With the help of Russ Marshall's company, Kappa Scientific in Santa Barbara, Dr. Ralph Wuerker developed in 1962 with Laikin's help a Kerr Cell Q-switch filled with a carcinogenic liquid, nitrobenzene. It quickly went obsolete and was replaced by a Pockels Cell, containing a solid crystal. In addition to being safer, another big advantage was Milt Laikin designed it such that it could be produced in-house. This cut costs and increased profits.

Maiman's company in 1962 was the first to offer a Q-switch far superior to the clunky rotating mirror Q-switches, called "mechanical".

---

[38] For more information on Q-switches – go to Appendix I where there's an explanation of the three main Q-switches: mechanical, electro-optical (Kerr and Pockels – both active Q-switches), and passive.

These were mainly offered by competitor TRG in New York. The Kerr Cell type produced far more power in a laser pulse and when the pulse occurred could be controlled. But Laikin's replacement, the Pockels Cell, was rapidly copied by the competition even though they could not buy it from Korad. Patents at Korad were not done at that time as Don Roberts covered earlier.

Thank You for Your Contribution—You're Fired!

Laikin shed some light on Maiman's thinking process as the latter assembled his brain trust at Korad. "His philosophy was that engineers and people of that sort have their greatest ideas when they first come to a new environment. So therefore, I and all the other engineers were expected to have very high productivity the first few years we work there. After that, we would have no more n*ew ideas*. So therefore, it would be time to get rid of us. That was his philosophy."

This particular observation was similar to my experience while working in marketing for Korad. Fridays were frequently referred to as "Greenie Day," when people got fired and were given their green slips. This likely explained the high turnover rate at Korad. Laikin continues: "One time I made a very large Kerr cell Q-switch; it was really a thing of beauty. And Mueller ground and polished all the surfaces, then I filled it and we had to clean it. We put it into a degreaser and the guy who worked for me took it out too fast and the damn thing cracked. When Maiman found out about that, he fired him, which wasn't right. I was tempted to walk out."

Laikin's observation about Maiman's habit of firing people after they had, in his view, expended the bulk of their creativity and potential, was reinforced when he himself was dismissed after a similarly brief period. "Maiman felt that I had already developed all the things that could come out of me, and that was it." Laikin's assessment of Maiman as a person contrasted starkly with his assessment of him as a scientist: "Maiman, I very much admired. I thought he was very, very bright. But he didn't care about people as human beings."

Sales manager Marvin Sachse agreed with Laikin's assessment. He related in our interview of March 3, 2009, how Maiman got very upset with Sachse for having a good-bye party for the employees at a company, Idak, founded by Maiman after Korad. Sachse was even willing to pay for the party out of his own pocket, but, as he told me,

"Maiman just wanted to close the doors and not thank the staff for their efforts."

And Then, You See, There Was This Little Issue of...

Some additional perspective on the possible motivation behind the short tenure of many at Korad was revealed by Ben Parks, one of the early Korad crew, in an interview conducted at his picturesque home in Northern California's gold mining country, Placerville. At seventy-two, Parks lives with his wife, Gayle, and his appearance, during our interview of March 5, 2010, was in harmony with his rustic surroundings, the main feature being a brightly colored, hand-woven Ethiopian hat. Parks had been selected by Maiman because he was a promising electronics expert who had been pursuing a doctorate in plasma physics at UCLA.

"They didn't want people to vest[39] in the Union Carbide retirement plan. So a lot of people got the axe at four years, eleven months, including me. That was strictly a Union Carbide policy, to get rid of people before they vested, I think. Toward the end of the five-year period, they started firing people and it was very upsetting." However, Ed Young, Maiman's financial controller, told me this was not Carbide policy. So it was perhaps Maiman chopping out his version of dead wood, those who had been milked dry of useful ideas. From a "brutal" business point of view, he was entirely correct in chopping out dead wood, like a gardener pruning her roses for the sake of overall improvement. From my years as a company president in the field of lasers, I have to admit it goes with the job. You aren't running a welfare outfit. Maiman would have been a good gardener; for sure he was an excellent businessman, as Ed Young strongly emphasized in our interview.

Unlike Laikin, however, Parks harbored no hard feelings for his former boss. "I never had any problems with the man. I thought he was a courageous person. He was a hustler. He got out and he did things without maybe looking around too much. He won the race to create the first laser because he was creative and he worked at it. Korad was a fun and wondrous place to work. We were blessed to be a part of it."

---

[39] Vesting means the employee would be entitled to retirement rights.

As Korad flourished, Maiman's brilliance as a scientist and businessman ultimately came to overshadow his deficiencies as a negotiator and diplomat. The company was busily pursuing production of the Q-switched laser, the heart of its immensely high power product line for the ensuing decade. These models were produced in response to Carbide's directive that its government-funded projects should no longer comprise the full extent of its activities. Commercial applications were sure to result in lucrative opportunities as well, and Maiman's group was tasked with pursuing that fresh avenue.

Early and Important Innovations, Q-switches

More discussion of the remarkable nature of the Q-switched laser is appropriate here, for these devices were amazing for their time. They were capable of producing incredibly powerful laser pulses that lasted for such a tiny instant as to defy appreciation. Contrasted with the extremely low-power continuous-beam communications lasers being produced by Bell Labs and others for use in communications, the enormously powerful (only when Q-switched) pulsed laser delivered the laser radiation in short bursts, an entirely different animal with an entirely different set of potential applications. Most of the laser uses described in this book are produced by Q-switching, including the first billion-watt laser (chapter 4), Moon ranging (chapter 5), and smart weapons (chapter 6). It may be of interest to note that laser welding never involves a Q-switch; when welding you want to melt, not vaporize.

"Q", by the way, is for quality of energy storage. A laser stores light energy between its mirrors like a child on a swing stores mechanical energy. As the swing goes higher, more energy is stored. Get in front of the kid, and the energy gets dumped on you. A Q-switched laser dumps stored light energy in a powerful, short-duration, giant pulse suitable for vaporizing things and for laser (radar) range finders.

Laser cavity is an extremely important concept in understanding all lasers. It is what is between the two end mirrors that make the light oscillate (bounce back and forth), getting amplified on each pass through the laser head that sits between the mirrors.[40] All Q-switches, except the mechanical ones, lie between the two end mirrors. One

[40]How lasing always gets started is that it is preceded by spontaneous emission in the form of fluorescence which quickly converts to stimulated emission.

mirror, called the rear mirror, reflects almost all of the light. The front mirror lets some of the light go through it. This is the beam you use. What gets reflected back is amplified in the laser head. Then it reflects off the rear mirror, back and forth, lots of bounces (oscillations). You might visualize the particles of light, the photons, as rubber balls bouncing back and forth. Or equivalently, you can picture waves. When the Q-switch opens, it switches the light stored in the laser head out of the front partially reflective mirror. As we say in the laser biz, it's all done with mirrors!

Bernard (Bernie) Soffer, 1967
Developer of the Passive Q-switch[41]

The Discovery of Mode Locked Pulses

I must digress to tell my favorite Maiman story that has to do with an odd characteristic of the passive Q-switch, now called mode locking. Within the passively Q-switched giant pulse there were actually not one, but many sub-nanosecond pulses called picosecond pulses. We knew these fast picosecond pulses, thousandths of a billionth of a second, to be exceedingly high power and therefore troublesome. We figured they were the cause of "bubbling" in our ruby rods, a string of fractures that looked like bubbles running down the center of the rod and usually close to the exit end where the power would be the highest. Naturally, bubbles were permanent and ruined the expensive rod, making it junk.

[41] For more on the passive Q-switch read Appendix I.

One day Maiman came into the lab where I was testing high-power eight-inch-long rubies in a K-2 laser, big brother to the K-1 whose rods were half the size of the K2. He was visibly worried about zapping so many highly expensive rubies ($3,000 a whack) and had heard on the grapevine that I was testing a solution. The idea floating around amongst the technical staff was that by "de-tuning" the laser by slightly misaligning the end mirror in the laser cavity, we could get rid of the pesky picosecond pulses. I had just verified this theory on our fabulous Tektronix 519 scope, and Maiman wanted to see it with his own eyes. Afterward, he grunted and walked out; remember he was not one to talk much, especially to me, a lowly engineer contaminated by working half time in marketing. Besides, I knew from my marketing boss, Tony Johnson, that Maiman wanted to get rid of me, but he, Johnson, talked him out of it.

What this story illustrates was Maiman tending to be more of a businessman than a scientist at this point in his life. He missed out on a glorious scientific opportunity to explain mode-locked picosecond pulsing, which shortly after was accomplished at Bell Labs in Murray Hill, New Jersey.[42]

Multi-layer Tough Laser Coatings

Other innovations flowed from the labs at Korad, including the replacement of fragile, silver-coated laser end mirrors by long-lasting coatings of magnesium fluoride and cerium dioxide. These tough coatings were applied to mirrors, windows, and all laser optics that the new powerful ruby laser beams were striking. As the innovative optics engineer Laikin explained: "This was known as a dielectric coating to distinguish it from the metallics like silver that preceded it. Some of the coatings that we were doing had 20 to 50 layers, so we put on the magnesium fluoride, cerium dioxide, and so forth, to stack up the coating layers."

These coatings were very thin; a fraction of a laser's wavelength, nanotechnology way back then! By this method Korad could get any reflectivity they wanted from low to high. The alternating layers of dielectric material could produce partial reflectivity of any percentage

---

[42] M. A. Duguay, L.L. Shapiro, and P.M. Rentzepis, *Physical Review Letters* (Oct. 30, 1967).

desired. By having fewer layers of coatings, the partial reflectors were used (and still are!) as the output mirrors in the laser cavity (where the beam came out of the laser). The ones with a lot of layers were the rear mirrors.[43]

Of course Korad had their very own optics shop to apply these hardy coatings. Needless to say, this gave the company a hefty leg up on the competition. As Maiman noted in *Odyssey*, Korad dominated the world throughout the 1960s in selling the world's most powerful lasers. But these were scientific lasers. Beginning in 1967 with my promotion to head up the newly formed industrial sales and marketing, called materials processing, the company's emphasis began shifting to machine tools and away from military and scientific markets.

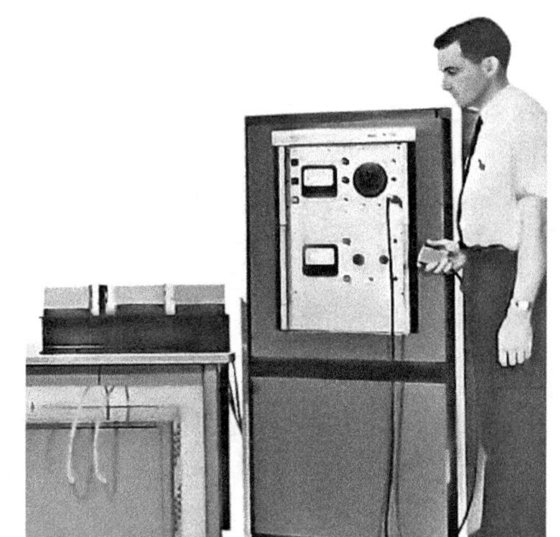

The author "firing" a Korad K-1Q[44]

Yes, we certainly kept busy during that period. One fascinating development seemed to follow another, and our activities were followed avidly by our East Coast competitors as theirs were by us. Little did we realize at the time, however, how much curiosity and interest our developments were arousing in certain other quarters.

---

[43] This is the essence of how all practical lasers work, as we say in the laser biz, It's all done with mirrors!

[44] Figure 8, *Laser Focus* (Oct 1965): 18.

Chapter Four: Spying and the First Billion-Watt Laser

Over time, these new lasers revealed their potential to provide solutions to manifold scientific challenges and became the subject of considerable Cold War curiosity on the part of the Soviet Union, which characteristically attempted to pry and cajole as much information about them as possible from Maiman as well as many others in the field.

The Russians Eye Korad's Activities

Maiman, perhaps naively, felt the American scientific community of that period was overreacting when it responded coldly to the overtures of Nobel Laureate Alexander M. Prokhorov during a visit by the latter to Southern California to attend an international laser conference. Maiman subsequently invited Prokhorov to visit Korad for a sanitized tour of the facility. He found the Russian to be quite engaging and extremely curious about Maiman's activities, but not the least bit forthcoming about his own research. Maiman reported in his autobiography being circumspect and cautious on these occasions, although he noted that some of his fellow scientists may have been somewhat less discreet during Russian-hosted dinner parties where "the vodka flowed generously."

His interactions with Prokhorov did not go entirely unnoticed by the CIA, who subsequently met with Maiman, suggesting that he keep his eyes and ears open during ensuing visits and that he report back his findings. This early period of the laser's development coincided with the high point of the Cold War, and some other anecdotes of a similar nature, provided by others at Korad, include details more reminiscent of the period's beach-blanket movies rather than anything particularly nefarious.

A Trip to Communist East Germany

Ben Parks, Korad's electronics expert, admitted that he never finished his doctorate in plasma physics. "It was very hard work. Plasma physics is a very serious career, a lot more than what I wanted to learn about. I was enjoying life at Korad, and I was enjoying going surfing on the coast. Life in California was good."

Parks commented on a piece of advice he received from Korad's party-pooper "security guy," Don Kopczick:[45] "I had a lot of credit time to burn off because we worked twenty hours a day, so I had enough time to go canoeing in Europe. Kopczick called me and said, 'Ben, when you go to Germany, you are going to meet all these Communists. They are going to try to get information out of you. They'll take you to their house, they'll give you alcohol and women, and then they'll pump you for information.'

As it happened, we arrived at the Berlin Airport, but the people who were delivering our kayak boats were not there. I called England and they said, 'Oh no, you have to come to England to get your boats. We're not delivering to Communist East Germany; you have to come here.' My friend and I didn't budget for a trip to England, so we decided to hitchhike from Berlin to London. The first ride we caught was in a Volkswagen bus. The driver turned out to be none other than the president of the Green Party of East Germany. He took us to his home where there was a Green Party rally in full swing. There was a lot of drinking, dancing, and partying. I was thinking to myself: *I've only been in Germany two hours and I'm already in this nest of Commies.*"

---

[45] Don Kopczick was the generally disliked personnel manager (the position now called manager of human resources). He was in charge of hiring and firing. Like most of us, he was hard-working and got results, such as when he interviewed in various US cities for an engineering manager and found Jim Boyden, a brilliant choice.

Missteps Amid the Rush to be the Best

The Russians maintained a distinct determination to acquire as much information as possible about US progress in the various fields of science. This environment in which our brightest people were not only competing with their colleagues in the field but also with those abroad led to any number of missteps, mostly the result of pressure to be the first, the fastest, the best.

One particular stumble was related to me by Dr. Jim Boyden, Korad's extremely inventive physicist/engineer, engineering manager from 1963 until 1968. Boyden, a boyish-looking man, dynamic and animated in his conversation, was hired by Maiman because of his army experience in laser research at Frankford Arsenal, Philadelphia. The army was interested in lasers as potential target designators or range finders. He had built a ruby laser with only Maiman's publication to guide him, a remarkable feat in the early 1960's.

Boyden's credentials were typical of those assembled by Maiman in those early days: the cream of the crop. Sporting a master's in physics from Carnegie-Mellon—called Carnegie Tech when Boyden received his degree in 1956—as well as a handsome fellowship, he headed west and earned his doctorate in physics from Caltech in 1960. Now seventy-six years old and recalling these events in an interview conducted in February of 2010 at his bachelor mansion in Ojai, California, Boyden detailed the difficulties involved in meeting the requirements of every client.

James (Jim) Boyden

Pioneer Gets Arrow in Back

In 1964 Korad had a contract with a British government nuclear lab, Culham Laboratories[46], to furnish a laser more powerful than any that had ever been built. The contract called for a billion watt laser, the world's first. It is also called a gigawatt (GW) laser. Boyden said, "Unfortunately, much to our surprise, it self-destructed. When we discovered this damage, we had to back off on the power. We couldn't deliver the spec because it was killing itself, so we had to lower it. We had to reduce the power because we had (the GW) spec to meet and yet we could increase the brightness. We were trying to argue that we still met spec because the brightness was what they needed for their work.

There was a lawsuit between Union Carbide and Culham Labs. They said we didn't deliver to spec and they wouldn't pay us. We had a Carbide lawyer with us in the UK. Even though we had delivered the brightest laser in the world, it didn't matter because we didn't meet the peak power spec in the contract. However, it was the application of that level of power that caused the ruby damage, so we could not meet this spec. We had submitted a bid to Culham on their pulsed peak power spec before anybody in the world had discovered this damage mechanism in high-power lasers." So Korad the pioneer got an arrow in their back!

The heart of this equipment, the synthetic ruby crystal, which, when stimulated, released its energy in the form of a laser beam, was a very expensive component. It required considerable caution and care when transported to the client's location.

Boyden continues: "I remember before we shipped the laser with all the multiple amplifier laser heads and spare rods, we took all the rods out because we didn't want to ship the laser with the valuable ruby rods in it. I carried them all in my suitcase, all of these big laser rods. When I was going through customs in England upon arrival, they asked me, 'Do you have anything to declare?' I said, 'Oh, I have 340,000 carats of ruby with me.' They laughed and said, 'Go away.' So I did. These hundreds of thousands of carats of ruby weighed several pounds. While working on that laser system I visited England two or three times, including once when we had their lawyers involved.

---

[46] "Culham", England's premier lab where efforts in nuclear fusion are carried out.

Carbide sliced the price in order to settle. Here they were slicing tens of thousands of dollars off my beautiful laser. It was very disappointing."

Are We There Yet?

The various household applications of the laser that prevail today, from CD players to laser pointers, laser printers, CD ROM drives, and the like, would seem ridiculously unlikely to those who insisted in the early sixties that the laser would never amount to anything other than a clever novelty. The initial stage of the laser's gestation met with a comically impatient response from the media of that day, which petulantly demanded immediate solutions for every conceivable scientific challenge. Having hyped the laser in order to sell their product, various journals became frantic when their initial expectations were not immediately gratified. "Do you think the laser will ever be practical enough to be used in anyone's home?" a trade journal editor needled Maiman (*Odyssey* Pg. 166).

This question is reminiscent of the query heard by every parent during the family vacation: Are we there yet? And indeed, there were some pretty spectacular accomplishments on that metaphorical trip to Disneyland, the journey to today's plethora of laser applications. Two particularly fascinating examples stand out: The resourceful efforts to remove carbon inclusions from otherwise hugely valuable gemstones (an undertaking requested by a diamond merchant to enhance their value) and the Apollo-era extraordinarily precise calculation of the distance from the Earth to the Moon.

Korad's involvement in these accomplishments was to take place in the absence of its founder, however, as 1967 was to be Theodore Maiman's last year with the company. Over the years Carbide had been buying out Maiman's ownership and also the Maiman group. By 1967 they owned Korad completely, one hundred percent.

1Gigawatt in 5 Nanoseconds

With Korad's new K-1600 Ruby Laser System

This Laser system uses state-of-the-art Techniques to produce narrow laser pulses for applications requiring improved time resolution, such as rapid plasma density fluctuation studies or high accuracy ranging measurements. The 5 nanosecond pulsewith is achieved by use of the pulse transmission mode PTM) oscillator technique; high loss pulse chopping techniques are not required. This oscillator-amplifier system thus produces high brightness concurrent with long ruby life. Beam divergence is typically 1.0 milliradian (1/2 angle; 1/2 power).
For further information on how the K-1600 can be of value to you in your work call (213) 393-6737, Ext. 292, or write Korad's Sales Manager.

Laser Focus magazine ad, Dec 1968, pg. 13
(a gigawatt is a billion watts)

Chapter Five: The Hal Walker Story

How a Black Man Beat the Cards Stacked Against Him

Hildreth (Hal) Walker 1966

It was in early 1968 that I decided to try to get the highly competent Hal Walker promoted as my assistant. He was tall, personable, and the leader of a rhythm-and-blues band that played until dawn many nights, yet he never showed any signs of fatigue the day after. His stamina was important to me as my job involved travel throughout the United States to install lasers at universities, companies, and government labs. The work was challenging, as the customers were professors and/or scientists and many had doctorates in physics. I thought perhaps they would be a tough bunch to deal with, but when it came to lasers, they were an attentive lot and eager to learn. The laser had been invented less than ten years prior, so for many of them, these were the first they had seen. The ability to teach professors and scientists about lasers while simultaneously working with your hands was a dual skill not shared by many of the working staff at Korad. Walker and I were the chosen ones to travel, install lasers, and help the sales representatives on sales calls. I wanted, however, to get out of this job which I considered a dead end. I feared being stuck in this capacity forever unless I could get someone else. I was certain Walker would be a great fit.

Korad had many physicists who were capable of doing my job, including Jim Linn, Bill Rundle, and Don Smart. Then there were Drs. Jim Boyden and Bill Buchman, both of CalTech. Alas, I always did a good job and produced satisfied customers who were happy to pay. As a result, I was a victim of my own competence—or at least that's how I viewed the situation.

From the vantage point of old age, I now believe Walker and I merely took pressure off the technical staff, allowing them to do their jobs without wasting their valuable time in traveling. Of course, getting customers to pay was critically important—something akin to selling. I knew Walker to be an excellent choice to replace me, not only because of his stamina as a late-night musician and band leader, but also because of his charisma; I thought he'd be adept at charming the physicists who comprised the bulk of our customer base. After all, he had completely charmed practically everybody at Korad, from Maiman on down.[47]

Hal's job as a laser technician required him to check out the lasers and do final testing just before they were shipped. He would fill out a form listing the measurements he had made. He had to make sure the laser performed to the customer's specifications. The most important thing was meeting the specs; if they were met, the customer was obliged to pay.

These were the early days of the laser. We had a saying at Korad: *Put it in a metal box, paint it blue, and ship it!* What this meant was as

---

[47] I asked Hal to explain his charm secret. He replied, "Simple—I just talk to them about their families and home life." So then I knew why I failed in this department; I was all business.

soon as a new laser worked in the lab, find a customer and ship it. Although our jobs were far from easy, they were challenging and paid well. When I returned from installing a laser, I always made it a practice to brag about exceeding even Hal's numbers whenever such was the case. "I even got more power out of that laser than Hal Walker did," I would boast.

I would repeat that message to anyone who asked me how my trip went, including Korad's manufacturing manager, Hal Moss; his second in command, Herb Stein; or my boss, Tony Johnson, the marketing manager. Even Dr. Fred Burns, Maiman's second in command,[48] received an occasional update from me regarding Walker's valuable contributions. I never mentioned it to Maiman, however, as I was too far down the ladder for him. Burns was the operations manager and that position carried considerable influence. In scampering about and extolling Walker's professional virtues I had a hidden agenda: to hammer into everyone's mind that a black man could excel in a position that they felt only a white man could manage. I knew that to get myself promoted, I'd have to find a replacement who was the same or better than me. Walker was the man.

Previously, I had been allowed to hire such a person: Don Tate, who unfortunately had a wooden leg. Don had assured me this wouldn't keep him from traveling. However, after I hired him, he refused to travel, but instead made himself so useful writing the instruction manuals (which was part of my job) that I couldn't fire him. So I was stuck in my job and unable to hire another. Walker was my savior. I was desperate, but because of the Tate mistake, I couldn't chance a second one. Walker was simply the best man for my job; there was nobody else.

I was encouraged when Johnson, our boss, told us his sales manager, Marvin Sachse, had advised him that he (Sachse) was considering promoting Walker to sales. "He has a great personality. Look at how he has all our PhDs eating out of his hand," Sachse had said. "They love to teach him all sorts of stuff about our lasers. I'm jealous; they don't do that for me. They can't wait until I leave their offices. I think he would be great in selling to our PhD customers." Johnson promised to consider the matter, which meant that he'd have to run it past Drs. Maiman and Burns.

---

[48] Drs. Maiman and Burns were next-door neighbors in Pacific Palisades.

Marvin (Marv) Sachse, 1970

After receiving his marching orders, the next day Johnson entered Sachse's office and said, "Well, I've thought about Walker in sales and decided that even though he can do the job, he's colored, and therefore not acceptable. Sorry, that's the way it is." Johnson had a way of tightening his thin lips that announced he had made up his mind. That and his piercing eyes framed by his glasses and sharp nose gave him a hawkish appearance.

Sachse seemed to agree, replying, "I agree that Walker in sales might be a little controversial and not be in Korad's best interests. People can say no to a *white* salesman but they might have a hard time saying no to a *colored* salesman. How about the customer service department? People with a problem to solve should tend to be less prejudicial."

"Got a point there; I agree," Johnson responded, taking a puff on his cigarette and talking fast as he always did. "But Waters is doing a good job, so there are no openings in customer service. Besides, we just hired Don Tate as his helper to write and take the load off his back on those instruction manuals. That gives Rod more time for sales support, making sales calls with the reps, and saving the company money by doing double duty when he's out installing lasers. Let's just keep Walker in mind for the future."

Sachse later told me he initially hated himself for agreeing with Johnson because it made him feel like a bigot. He said he and Walker were friends, which further contributed to his discomfort. In retrospect, he concluded that he was vindicated. It was the correct decision,

Sachse rationalized. After all, Walker would no doubt profit far more from being a techie than a sales guy, right?

Walker did not have a college degree at that time, but he was the company's ace when it came to the final checkout of laser equipment. I continued pressing my agenda on his behalf, continuing to tell everyone that I had gotten even more out of a laser at the customer's location than Walker's impressive performance had provided back at the company.

When I judged the time was ripe, I approached my boss, Johnson, saying, "I feel I can be more useful to the company by devoting full time to sales of spares and accessories." These highly profitable laser spares and accessories brought in a consistent $70,000[49] every month that required no selling—order taking as Johnson called it—the same as a sales clerk behind a counter.

"Hmmm," Johnson responded. "Sachse and I have been discussing a promotion for Walker. Let me think about it some more. But, you know, it's unheard of for Negros to be lead men in the machine shop, much less supervisors and certainly not managers. Their being a manager is totally unheard of and risky for our reputation."

Although less than confident that Walker would be given this opportunity, I was excited and encouraged that my efforts might still pay off. Not only would this promotion be a boost for Hal, but it had the added benefit of serving as my ticket out of customer service and all the traveling that position entailed. Having a wife and three small children, I was looking for a position permitting me to be home more regularly. I was disappointed when Johnson came back to me saying, "Sorry, Hal Walker is too valuable in his job. Hal Moss won't let him go."

Not one to give up, I kept at the company and its management for a total of nine months. One of my little laser accessory items, a calorimeter used to measure the energy in a laser pulse, was selling well. Back in the lab I had figured out how to eliminate expensive calibration using cheap light sources instead of costly working lasers. Here is an amusing Maiman story illustrating this: He asked me how I thought replacing coherent lasers by an incoherent light source was justified. "Photons are photons, I guess, whether laser or

---

[49] With inflation Korad's spares & accessories business works out to be $480,000 in 2013 every month—almost $6 million a year!

conventional," I replied. He frowned, shrugged, and walked away without comment, probably thinking, *wise guy.*[50]

The rising income being realized from the sale of my parts and accessories made management consider the possibility that I might be more valuable selling laser machines. Fred Burns had apparently been suggesting something to this effect, so I once again proposed that Walker be promoted to replace me at my current position. "Sorry," Burns replied with another puff on his ever present cigar, "he is too valuable in his job, and besides, a Negro can't go to the South to install lasers."

To Savannah and Back

Knowing that this was a company of physicists who loved experimentation, I replied, "How about sending Walker and me to Savannah, Georgia, to install the Department of Agriculture's new ruby laser? Then if we both come back alive, he gets the job. OK?" A slightly twisted smile edged its way across Burns' face, indicating some surprise at my audacity. I had learned from working at Korad for two years that audacity and humor could often win the day. They called it *chutzpah* and he was a master of it.

Before Walker and I left on our trip, I told him all of what Dr. Burns had said. There were no secrets between us. A cornerstone of our friendship was to *tell it like it is.* He and I flew to Atlanta and rented a sporty, yellow Mustang. We were now in the heart of the South during the height of the civil unrest that marked the 1960s. "I think it best that I drive," Hal intoned. "It will look better. It's OK for us to drive white folks."

As we drove off, my job was to keep a nervous eye out for rednecks, who in those days and in that place often violently objected to a black and a white man traveling together. Only once did we encounter rednecks. As we were looking for a place to eat lunch, two young white guys roared up, then hit their brakes to stare at us malevolently before stopping at a gas station. Outside a likely café hung a sign, "Whites Only." As it was the only eatery we had seen, I asked the hostess if we could possibly dine there. It was after the lunch hour and there weren't many patrons. She was older and probably the owner. We were admitted. Shortly thereafter the two

---

[50] Photon is the name for a particle of light.

rednecks came in, once again glaring at us. We could see them talking to the hostess, pointing at us all the while. As we were in a hurry to see our customer, we had requested fast service and were already eating. Obviously, the owner didn't want to evict us, so we hurriedly finished and left. Walker yelled, "Quick, run to the car! Let's get out of here. I'll drive."

Jumping in the yellow Mustang, we blasted off as the rednecks came running out. Thanks to Hal's idea of renting a high-power car, we lost them in their decrepit old car. Relieved, I said to Hal, "Good thing you're driving."

Later that day, we installed the K-15 laser[51] at the agricultural lab, got it operational, and then at night we hung out at a black night club where the only other white person, a woman, spent some time glaring at me. I supposed at the time she felt I didn't belong there, but I wasn't interested in either her or her reason for shooting daggers at me. I was not there to give offense or be uptight but simply to have a good time. Besides, the other patrons paid me no notice, or so it seemed to me. I marveled at Walker's socializing abilities as he quickly made friends left and right. Everyone was having a great time, dancing briskly to the rock 'n' roll. I had a wonderful time carousing with this new group of friends. Besides, I got to see a side of Hal I had never seen before; he had such charisma and was such a performer—even without his musical instrument. I knew for certain he was the man for my job—and then some.

If I may digress a bit on a favorite subject, Hal . . . he made an enormous impact on many people. With his wife, Bettye, a PhD educator, he founded A-MAN, a company that encouraged ten-year-old black kids to go into science. Over the years they have racked up hundreds to possibly thousands of converts to science careers, not only in the USA but in the other USA, the Union of South Africa, where he owns a second home and escapes the Southern California winters. After the Cold War ended in the early 1990s, he sold hi-tech Russian Faraday Rotators to the American controlled nuclear fusion folks.

There's lots more, but you get the drift. And one more thing, after

---

[51] Model K-15 was a fast version of the venerable K1, the gold standard of lasers in the 1960s. The K-1 was also the oscillator laser head in a K-1500 oscillator-amplifier, the world's first billion-watt laser which sold for $40,000 in 1965, worth $300,000 in 2013.

Korad, he went to night school and got a degree as a requirement for promotion to middle management at Hughes.

Now we return to our story . . . Dr. Burns might well have been disappointed that our Georgia trip had elicited so little outrage, for he was still reluctant. "It just isn't done, having a Negro in a responsible job. What would people think?"

Personally, being a lifelong admirer of the Jewish people, I attribute Walker's rare and outstanding success in landing a top job in the white man's world to Korad being managed by Jewish men who gave everyone the day off with pay for the High Holy Day of Yom Kippur. I believe that my getting Walker away from manufacturing was because, in an almost earth-shaking meeting with Burns, Sachse, and Hal Moss, I said, "You guys wouldn't discriminate against a poor black man, would you?" Sachse looked at Moss and their eyes locked for a moment—then they both shrugged.

Fred P. Burns

Harold (Hal) Moss

I could see the shock waves rolling across Fred's face and heard the nervous laugh from Hal Moss. That broke up the meeting. I had won. Apparently Dr. Burns finally realized he was tethered to a sinking ship and resigned himself with a huff and another puff on his cigar. "OK, he's yours. Now go find something else to do!"

I immediately went to Walker, telling him about the meeting and saying, "I didn't tell them this, but what I really want is for you to take my job." I remember quite clearly his eyes opening wide with surprise and obvious disbelief. I hastened to convince him that I was serious. I told him I wanted a promotion to product manager so that I could be in charge of selling new products. "Now that you have your foot in the door as my assistant, it's up to you to get your whole body in and take over as customer service manager."

In subsequent years, his outstanding performance in this new capacity put to bed at Korad all the hysterical nonsense about black people being incapable of leadership and holding down a white man's job.

How Walker Had Landed the Korad Job

Walker had begun at Korad as a contract worker in 1964, just four years prior to this point. His experience in the BMEWS system during the late 1950s and early '60s had provided valuable experience. BMEWS was an acronym for Ballistic Missile Early Warning Systems, a component of our Cold War strategy near the US border with Soviet Russia in the Alaskan wilds, near where the dreaded Soviet nuclear missiles were expected to originate. Walker had noticed an ad in the *Los Angeles Times* specifically requesting applications from those with experience in heavy ground radar. High-voltage systems were being utilized at Korad, the same type as big radars. Because carelessness could be fatal, experience with handling large-energy discharge systems was very necessary for Korad's high-power lasers, which were the world's most powerful. Already an electrocution death on a laser had occurred elsewhere; the danger was real.

The time of this story was at the tail end of the Jim Crow[52] era; it was a white man's world in those days, and for people of color the door of employment opportunity had opened only a mere crack. Hal recalled his apprehension before, during, and after his early 1964 Korad interview with Bob Grossman: "Naturally, I had concerns because I wanted that job very much. I was really quite impressed with lasers and what lasers could possibly mean as a technology. So I was expressing myself as broadly as I could to get his attention and to let him know that I could do the job. I was frightened a little bit about the fact that I was an African American. I was worried I might be turned away because of my race. This had happened on other

---

[52] Jim Crow laws, as they were called, legally discriminated against African Americans only. The Jim Crow era went a full century past the American Civil War up to 1964 with the passage of President Lyndon Johnson's Civil Rights Act. Note that our story occurred five years later.

occasions in the past, so apprehension was for good reason. Much to my joy, I was accepted and, of course, I was thrilled and eager to go to work in this exciting new area of laser technology."

Walker's concerns were well founded, given the atmosphere of the day, but Korad was not your typical company; those in charge generally thought out of the box, and were quicker than the general public to modify their assumptions where race was concerned. Because we had Asians, African Americans, Hispanics, and Europeans, we Koraders used to call ourselves "International Lasers." We were proud of our racial diversity.

Walker's hiring was part and parcel of an occasionally grudging but generally good-hearted evolution in the company's thinking, an evolution taking place nationwide, which reflected Dr. King's call that folks be judged on the content of their character and not on the color of their skin. At Korad, the emerging consensus was that the relative worth of a given employment candidate should be considered without regard for their race.

The Best Policy

Walker found Grossman to be a tough interviewer, discerning, and in no manner patronizing. Grossman was fair but unfaltering in his expectation of honesty. Walker recalled: "He was quite candid that the work that was going to be done there had a lot of risks involved because we were working with high-energy discharge capacitors and electrical powers, probably many above 10 to 15 kilowatts. These could and did kill people. I told him that I had handled systems like that before, but I was not familiar with high-energy discharge capacitors."

"He asked Walker a very interesting question: "Hal, tell me what you know about the word *joule*." Walker thought about it for a moment as he was familiar with the term. But he had never used it in a technical discussion so he decided to reply, "Bob, I don't know about joules." Then Grossman happily explained it as it applied to laser energy measurements. But he commented very quickly that he was quite pleased Walker had told the truth. He said many people he interviewed had tried to make up stories about what joules are all about. He detested that type of an individual who would not be honest about what he knew and did not know. Walker told me, "Based on that one fact that I was truthful, he told me, 'I'm going to hire you to work here.' Period!"

And thus another hurdle had been cleared; not just in the sense of a civil rights obstacle overcome, but as a personal achievement for Walker as an individual. He was to continue to be the exception to the rule for some time to come, for it was regarded as dangerous to hire a minority. Often referred to as "those people," African Americans were regarded as trouble. Being black in a white man's world meant it was almost impossible to find rewarding technical jobs. Walker watched in frustration as two close friends of his, Gene Wilson and Earl McLendon, both graduate electrical engineers, could each find work only as a technician, a non-college-degree position.

During our discussions in preparation for this book, Walker reflected on the attitude of that era: "Many times whites in key power positions felt responsible to enforce those barriers. Because for them to let those barriers down would sometimes be viewed as a detriment with their coworkers, as in, *Why are you letting those people in the company? We don't want them,* or *They're not good enough; they are inferior.* We were living in a time where that was normal. I was fortunate to get through that barrier."[53]

Walker found a couple other African Americans in the company's employ, albeit in manual-labor support capacities, such as crystal growing. Walker himself was given the title of manufacturing technician, later to become manufacturing supervisor and then manager.

Walker found the authority his superiors delegated to him at Korad to be very valuable when it came time to hire the right help. "We builders and the manufacturers of these systems had to be very accurate and very dedicated to our job and inspect our own work to a level that was much higher than I was familiar with in other jobs that I had had in electronics," Walker told me. "The type of people necessary to do this work, I found, were extraordinary and unique individuals. And when it came my turn to hire people into the company, I was very aware, from my own experience, what was needed."

These individuals took tremendous pride in their achievement when things went well and the installed laser worked as it should, as well as a grave sense of responsibility and determination when it did not. Of

[53] Some reviewers of this chapter have found these race-centered conversations to be repugnant. Hal Walker's reply is that there is absolutely nothing offensive to an African American in this book. (He laughed at the idea!)

course, the manufacturing department did not operate in a vacuum but could, and as necessary did, access the support of Korad's physicists like Jim Boyden, Bill Rundle, or Jim Linn, among others. It was a tightly coordinated and cooperative team effort, from the physical aspect of assembling the equipment itself to its testing and calibration, to the more technical questions involving optics, the extremely high-voltage power supplies, or something else beyond the scope of the fabrication process.

My Part in the Fight

To this day I am proud that circumstances allowed me to be a part of that team. So many years ago...it seems almost as in a dream, to remember the things we were involved with and the myriad applications that sprang from our pioneering work in laser technology. I take even more pride, however, in being one of those who rejected that hateful nonsense and ignorance that was responsible for the Jim Crow era.

Feeling a given way is one thing, but it's useless if we keep it to ourselves and never challenge the ignorant or assist the disenfranchised. To quote Albert Einstein: "The world is in greater peril from those who tolerate or encourage evil than it is from those who actively commit it."[54] When there arose an opportunity for Walker to be promoted and to directly interface with Korad's customers, I felt compelled to stand up to those who felt a black man wouldn't be able to handle the task. He and I recalled that opportunity in our 2010 interviews, which was appropriate, I thought, considering the first black president was in office. Listening to him recount the challenge he and his fellows faced, I was struck by the dogged tenacity, determination, and patience that must have been required of them.

He told me, "You approached me to inform me that I was going to be coming into the marketing department's customer service operations, and I had very strong feelings. Would I be able to do the job? Was I going to be prepared to take what I know out into an environment that I was unfamiliar with? But you reassured me and I remember the conversation we had about this.

---

[54] This was a favorite saying of Dr. Martin Luther King Jr., whose birthday is now a US national holiday.

"You assured me what I was doing before was basically what I was going to do out there except I would be doing it in somebody else's lab. I think what I was mostly attracted to was the opportunity to go out into the field and show—in sort of a selfish way—that black people can do this because we had been doing it: myself, Gene Wilson, McLendon, guys like that. We had been doing this for years only nobody outside knew it. It was inside the confines of our company.

"But here was an opportunity to go beyond that, and I was one of those African American people who always took an opportunity to try to exhibit possibilities for others to see that they can do this kind of thing too. You just have to take the leap of faith and go out there and work at it. But thank goodness for persons like you who took a lot of the fears I had about the job and said, 'Hey don't let that be the primary concern. Really believe that you can do it. Look at what you have been doing here. It's going to be fine.'"

That has to be the ultimate reward one receives when taking the right stand: the gratitude of those on whose behalf one stood up, earnestly expressed. While I was highly delighted to assist Hal Walker in stepping up to that challenge, I could hardly have expected what the job might have entailed, or what intrigue he might encounter....

Job Description: Espionage?

At one point in our interviews, Walker adjusted his glasses and shifted in his chair as he shook his head slowly, an amused grin lingering for a moment on his lips. His new position at Korad required some rather curious duties, he observed, bringing his gaze directly to mine. "It's a little tricky. The name of this story is espionage."

"The name is what?" I countered in curiosity and confusion as he smiled and shook his head again.

"Espionage," he said again. "You might remember a company in Indianapolis that bought some laser welders from us? There was an installation task there that I was going to work on. My flight to Indianapolis happened to stop over in Dayton, Ohio. Tony Johnson came to my office and said, 'Hal, when you fly into Dayton, we want you to get off the plane.' I was requested to get off the plane, go into the waiting room area, and go to a certain rental car company's counter. I was to have my briefcase with me, opened up. In the meantime a gentleman would walk over to me, put something inside the briefcase, and walk away. I was to then take that package on to

Indianapolis and then bring it back to Korad. At first, this sounded a little clandestine and I was apprehensive about this, but I was given no information other than to make the pickup.

"I did get to Dayton. I went into the airport and over to the rental car area where I opened my briefcase and stood there looking like I was doing something with the rental car people. Sure enough, a person did come up and stand next to me and put a brown envelope in my briefcase and then walk away. I never saw the man before—a complete stranger. I closed up the briefcase, got back on board the airplane, and continued on to Indianapolis.

"This bothered me sufficiently that I worried I was involved in something that had potential for big trouble. So while I was in Indianapolis, I never left the package alone. I was there for three or four days doing the installation. I carried the documents on my person because I didn't want to leave them in the hotel room just in case somebody was to 'discover' them.

"After a day or so I decided I had better look and see what this was because whatever it was, I was responsible. So I opened the envelope and removed the documents. They involved technologies using lasers and fire control systems for military projects." Walker leaned forward in his chair as he told me of the cryptic warning which was included on every page: *Possession of this page has a penalty, if convicted, of ten years in prison.* As he said, "Needless to say, I was freaked out!"

It was obvious at this point in our interview that it could be unwise to interrupt and spoil a good story, so I kept any clarifying questions to myself as he continued. "I can't remember how we talked about this later, once I returned to Korad, because I was very upset that I had been set up to do this with absolutely no knowledge of what it was about. I felt that part of my survival was due to my own, whatever you want to call it—*self-reliance*—to be able to keep that package protected, on my body at all times and not leave it sitting around on the desk or out of my possession.

"So when I returned to Korad, I went to one of the guys and I gave him the package. He thanked me very much, after which I told him that I was disappointed that they hadn't informed me that this was some type of espionage operation. I never let anyone know that I had opened the documents and had actually seen the information. I didn't think that was a smart thing to do. In those days, Dayton, Ohio, was home to the Wright Patterson Air Force Research Center. I learned later that the purpose of these documents was to give us a heads up on some particular military application that we were planning on proposing to the Air Force."

The Moon Laser
Ad in Laser Focus magazine, Dec 1969

Chapter Six: History-Making Lunar Ranging

Laser reflector left on Moon by the crew of Apollo 11.[55] **Note astronaut Buzz Aldrin's bootprints.**

Yes, the team at Korad Lasers was indeed resourceful, and the nature of the various challenges they faced was as varied as it was exciting. Certainly the highest profile scientific application involved the use of a Korad laser to obtain a precise measurement of the distance from the Earth to the Moon by bouncing a laser beam off

55 The retro-reflector consisted of 100 triangular-shaped silica corner cube prisms in a 10x10 array having the ability to reflect the laser beam back on itself so as to ensure a reflection back to the source at the California observatory where it could be detected. It's like hitting a racquetball into a corner and having it come back at you.

the reflector placed on the lunar surface by the Apollo 11 team of Neil Armstrong and Buzz Aldrin, the first men on the Moon and, indeed, the first humans on an extraterrestrial body.

There was intense competition for experiments to be done on this, the first mission to the Moon. The laser won for two main reasons: the importance of the science and the short time required for Buzz Aldrin to deploy the reflector.[56]

Maiman's rival, Charles Townes, in his autobiography,[57] begins with our same story of laser ranging. His choice of placing the Moon shot first is understandable because of all the great scientific discoveries that it made possible. One wonders, however, about his failure to credit Maiman's Korad. This stood out because elsewhere he takes pains to give credit where credit is due, especially to Bell Labs.

The lunar laser experiment was science of high importance in many fields: geophysics, experimental relativity, fundamental constants of the universe, and more. However, successful lunar laser ranging by the Americans was threatened by competition from the Cold War enemy, Soviet Russia. We simply could not waste the billions of dollars it took us to go to the Moon by having the Russians grab the glory.

The US was utterly determined to be first to laser radar off the Moon, so they made three lasers available to do the job. There were two at Lick Observatory near San Jose, California: one made by Spacerays[58] and one by Korad. Another Korad was at Haleakela Observatory on the island of Maui in Hawaii. There was a fourth laser, also Korad, but it was being built in Santa Monica for McDonald Observatory in West Texas.

Each laser at Lick—the Spacerays and the Korad—had a team running it. NASA ran Spaceray's and Wesleyan University ran Korad's. The competition between the two teams was intense and extraordinarily unfriendly, as we shall relate.

The Korad laser at Haleakala was built to track satellites in Earth orbits. It was not at all intended to reach the Moon and return. Nevertheless, NASA needed all the laser muscle they could apply. If

---

[56] *The Lunar Laser Ranging Experiment*, Bender, Currie, Dicke, et al, Science, Oct 19, 1973, vol. 182(4109), 229.

[57] *How the Laser Happened – Adventures of a Scientist* (Oxford University Press, 1999): 3

[58] Spacerays was one of Korad's many low cost competitors, located near Boston.

they failed, well then, the Russians would have a chance to grab the glory and a big propaganda victory.

During their two and a half hours on the Moon's surface the astronauts deployed three scientific experiments: the laser radar reflector, a seismometer to measure moonquakes, and a solar-wind experiment to catch particles coming from the sun. They also set up a video camera on a tripod for the viewing pleasure of millions of earthlings, then and forever. They took some historically memorable photographs, such as Aldrin's photo of his footprint and Armstrong's photo of Buzz with his image reflecting off Aldrin's visor, which, as he says, is "one of the most famous photos in history."[59]

Buzz Aldrin deployed the laser reflector on the Moon's barren surface. He and Armstrong then gathered rock samples that established the Moon was formed by a collision billions of years ago between two planets, one of which was our Earth. The Moon was knocked off the Earth, as conclusively proven by the moon rocks being of the same isotopes as the Earth.[60]

Distrust and Not-So-Friendly Competition at Lick

In the event of laser problems, NASA wanted the best man Korad had to be at Lick. This man was Hal Walker, now placed in charge of the Korad laser. Walker could not believe his employment at Korad Lasers had led him to this historic moment. This former jazz musician turned manager was now involved in a major endeavor that the whole world was watching.

Walker had finally succeeded in getting his laser in place in preparation for firing at the Moon. Time was getting short. The astronauts were scheduled to land on the Moon in two more weeks. Some considerable friction existed between the Wesleyan and NASA teams, understated as *competitive issues*. Of course, as we shall soon see, the term *considerable friction* is itself an understatement. The two teams argued over who was going to take the first shots at the Moon.

---

[59] The material in this paragraph came from Aldrin's book published in 2009, *Magnificent Desolation*. I had met him at an Explorer's Club meeting at the Marina del Rey City Club near the LAX Airport. For this book he recommended I read his 2009 book.

[60] The identical isotopic ratios on Earth and Moon are just one of many scientific facts that prove the American moon landings were real and not just a Hollywood stunt.

Predictably, NASA's Spacerays people felt they had the privilege, being *NASA* after all, and the ones running the Apollo show. As the days ticked by in the two weeks before the historic moon landing on the 20th of July, NASA put their Spacerays laser under the Lick telescope so they could score a solid historic first. But their laser quickly experienced a failure of its ruby crystal, an expensive and indispensable component of this type of laser. Walker was surprised when the NASA engineers asked him for assistance with their laser. Walker obligingly contacted Union Carbide, Korad's new 100% owner, and asked if they could provide some rubies. Unfortunately, when the crystals subsequently arrived and were installed, they did not work; the Spacerays didn't operate, and NASA was mad as hell at Walker and Union Carbide for having provided faulty rubies in what they perceived to be a scheme to give the Korad team an advantage. Now Walker and crew had the green light to replace Spacerays with Korad, and he was up to his knees in a blizzard of accusations and suspicion. But they did have a wonderful opportunity to make history.

Down But Not Out: Double Sabotage

The observatory's receiver of return beams from the Moon, designed specifically for this experiment by MIT, became the target of a saboteur. A jumper wire was placed across the fire control signal, which, working in conjunction with the receiver, could come into play only when the laser's tech fired the fifth shot. Walker's fifth pulse to the Moon set off the short which destroyed MIT's expensive and uniquely designed component. *Some people need a real good beating,* he thought as he resigned himself to the apparent reality that his team was now out of the competition, giving the Russians and the Korad laser at Haleakala the remaining shots at glorious victory.

The Korad laser at Lick Observatory actually belonged to Wesleyan University's Physics Department, who also had a former Korad physicist, Irv Winer, working there on his doctorate. Professor James E. (Jim) Faller at Wesleyan was contracted by NASA to come with Winer and his Korad laser from Wesleyan in Connecticut all the way across the United States to the Lick Observatory perched on Mount Hamilton just east of San Jose. Winer and Walker had worked together at Korad for a number of years.

Winer studied Walker with that odd, absentminded stare he got when contemplating a potential solution to a vexing challenge. His stare and round glasses gave him a decidedly owlish appearance. Walker knew his colleague's look well and took curious note as Winer characteristically slid his favorite stubby pencil from its constant perch atop his right ear and gripped the point between his thumb and forefinger, playing the eraser slowly across his lower lip. Winer was not the imposing sort—more along the lines of a slender bookkeeper than a hefty linebacker—but Walker knew this quirk all too well; his colleague was focusing his considerable intellect. And clearly he had an idea as he motioned Walker to accompany him and promptly turned and strode out of the observatory and into the bushes outside, where their conversation could not be overheard. Scratching a match across its box and putting the flame to a Winston, Winer huddled with Walker. "Let's go inside and act frustrated as though we don't have a plan B. Then we'll avail ourselves of one of the observatory's visible light receivers and modify it. I may be mistaken, but I believe I can make it work."[61] Several hours and one severely chewed eraser later, Winer had succeeded in analyzing the circuit design of Lick's appropriated component and placed it in lieu of the MIT gadget.

Within a day they were up and running and ready to face issues associated with the clearance in the surrounding vicinity inasmuch as the laser's calculated path was in the flight path of both the San Francisco International and San Jose Airports. The time available for firing at the Moon was consequently tight because both airports were busy. The prestige of the Wesleyan team likely would not be enhanced if, in the course of firing at the Moon, they blinded the pilot of a passenger jet, resulting in the deaths of hundreds. But they were able to arrange for a portable radar unit, courtesy of the local National Guard, and then had to wait for the all-clear signal before firing.

At long last they were ready to go with a replacement receiver. Walker felt his heart beating wildly with anticipation. At that moment the order came to fire, and he sent the first laser pulse to the Moon. His team members scrambled in an effort to keep the laser collimated in as parallel a beam as humanly possible. The objective was to get into position to track the Moon, which was wasting no time in its fast trek across the night sky. Walker fired a shot at the Moon when they heard a tremendous bang down below in the telescope compartment

---

[61] Winer's task was to maximize the receiver's sensitivity to ruby laser radiation at its wavelength of 694.3 billionths of a meter.

area. Walker whirled around as a wave of dread coursed over him. "Good God! What was that?"

Examining the laser's lower unit, he discovered that someone had created a short by bridging across two contacts with a jumper wire. He shook his head and shot a look of disgust and disappointment at Irv Winer as the acrid odor of a second sabotage permeated the room.

A Midnight Wild Ride Down Hamilton Mountain

Walker immediately called Korad, which was 350 miles down south in Santa Monica, and arranged for emergency repair of the ruined part. Loading the items in his rented car, he wasted no time heading for the nearby San Jose Airport. He headed down the looping, two-lane road connecting the observatory to the valley below. Built in 1887 as the main access road to facilitate the construction of Lick, a requirement was to keep it under a seven percent grade so that horses could pull their massive loads of observatory materials up the mountain. Consequently it twisted and turned back on itself many times, as horses are capable of turning on the proverbial dime. Hardly suitable for modern autos, it slithered slowly up Hamilton Mountain, offering its intrepid travelers a less than reassuring view of a sheer precipice at every switchback.

Walker was threading his way down the grade, looking forward to a nap on the plane to Los Angeles International Airport, LAX, when a set of headlights loomed ever closer in his rearview mirror. He had with him a laser coil box and some electronics that included the trigger circuit.[62] These had been damaged by the nasty arc that had cooked through the coil box, which he intended to take to Korad for repairs.

His eyes darted from the narrow roadway to his rearview mirror and back as he became aware that the car that had been following him for a few minutes was accelerating and gaining rapidly. After slowing to allow passage, only to have it pull to the side as well and positioned immediately behind his car, he remembers thinking, "I'll just stop and tell these guys, 'Hey, we're ranging to the Moon, and this is national science going on. You know, this is a great time in scientific history and we're doing good experiments. What are you guys *doing*? Why don't you leave me *alone*?'" Then they slammed into his auto, hard.

---

[62] The trigger circuit sent a pulse of electricity to the laser head that triggered (fired) the laser and produced a laser pulse.

Was this related to the nature of his work, or just a case of three rednecks out to show a black man what could happen when he ventured out alone? He couldn't be sure, of course, given the often hostile attitude expressed by whites toward blacks who dared to take on high-profile work and further dared to dress in suits, white shirts, and ties. As Walker continued with this story, it occurred to me that he was lucky to have survived the incident.

He related how terrifying it was as the vehicle then used his car as a brake; that is, his pursuers rammed his car and thus slowed themselves down at the approach of a curve, while simultaneously thrusting him forward. After a couple miles of turns at hair-raising speed and six separate such ramming maneuvers, it occurred to him that whoever was in the pursuing vehicle was unlikely to be persuaded by a calm, logical, and dispassionate approach. With the recent twin sabotages in mind, it was clear to him they might intend to kill him. He accelerated his sedan into the oncoming curve as hard as he felt he could, considering all curves on the road were unbanked and flat as boards.

"We just got to a curve where I was able to take advantage of my car's speed and stability, and they had to brake pretty hard because they couldn't make contact with my bumper; they had to hit their brakes hard, and I never saw them again." All that remained of his tormentors was a large cloud of dust, and he didn't hang around to determine whether they had merely spun out on the shoulder or left the roadway entirely. The consequences of either outcome, he decided, were theirs to suffer—one way or the other. Reporting the incident to the police was not an option for his ethnic group; he could face a murder charge if his assailants had been killed.

Korad did their duty, repaired the part and Walker was back at Lick with his laser now up and running again only 24 hours after the sabotage. And only a few days had gone by since the spacemen's lunar departure.

Finding the Laser Reflector

Walker had suspected that this moon project could require patience as most things tried for the first time bear out Murphy's Law: whatever can go wrong, will. But having already spent a month on the mountain preparing his equipment, he was more determined than ever to see how it stood up to the challenge.

The first twelve days of the laser-ranging operation after Armstrong and Aldrin's blast-off involved a search of their landing zone in an attempt to locate the lunar mirror. As a result, Korad's crews in California and Hawaii had received a directive to shift their laser's power level from nominal to maximum, widen the beam on the moon's surface, and begin a systematic search.

Both teams were feeling pressure from Washington DC to be the first to hit those reflectors by any means. This imperative resulted in considerable concern as it required a search over a broad area until they succeeded, requiring operation of the lasers at their highest capability for extended periods. It was feared that damage to the output surfaces of the laser amplifier might occur, and there was the potential for damage caused by overheating as a result of rapid firing of the laser.

Walker and the Wesleyan team had adjusted the laser beam so it would cover a bigger spot on the Moon in an effort to find the lost reflector. But spreading it out weakened the beam necessitating ratcheting up the degree of intensity and "overpower" the weak laser density on the Moon. Walker responded to the directive, running the laser at maximum power to cover more lunar ground in trying to find the reflector.

Friendly Competition Between Lick and Haleakala

The backup Korad laser, located atop Mount Haleakala in Hawaii, had been purchased by the Advanced Research Projects Agency (ARPA) and placed under the direction of its designer, Korad engineering physicist Bill Rundle, who had the utter misfortune of being deployed to the breathtakingly beautiful island of Maui, where he was tasked with modifying his ruby laser which was mounted piggyback style upon the observatory's thirty-inch telescope. Rundle shared Walker's thrill at being a part of the historical Apollo 11 project, even if he was stuck with a laser that had originally been designed to range off satellites, especially Russian. Range data was needed in the event a bad one had to be shot down!

Lick Observatory on California's Mt. Hamilton

Rundle had known Walker for several years and regarded his colleague warmly, but where history and glory were concerned, friendship would have to take second place to some good old competition. Rundle had been a busy man recently, designing and testing Korad's third lunar laser going to McDonald Observatory in Texas.

He had not seen his wife and children for a couple months and had shared his discontent on this point with NASA, who promptly insisted he have his entire family accompany him to Maui. Rundle's suggestion of a couple of weeks' sojourn, all expenses paid, amid the tropical beaches and lush forests of an island paradise met with little resistance from his family.

William (Bill) Rundle, 1969

What was NASA thinking, he wondered. Korad's laser setup on Haleakala Mountain had been created to illuminate enemy satellites for photography and to measure their distance. It lacked the power to reach the Moon. He had told them exactly that, but they insisted on trying anyway. As far as Rundle was concerned, this put him at a decided disadvantage to his good friend Hal Walker at Lick. "Damn, one shot at this and they saddle me with a close-range system," he mused. On the other hand, he and his family were on Maui while Walker had to settle for familiar terrain in California.

Rundle fiddled with his system design in order to increase its power, putting in a beam expander between the laser oscillator and the power amplifiers. He had occasion to break from his duties when distracted by the heart-stopping view from Maui's highest point. Haleakala was often quilted in fog at sunrise, with only its highest point, where the observatory was located, peeking out from the billowing cover. The sight of the surrounding ocean appearing as the cloud cover lifted in the warmth of the morning sun was something he would not soon forget. Coincidently, the mountain top, an extinct volcano, was devoid of life except for humans and a few plants; there were no animals. As such, its stark desolation had an eerie resemblance to the Moon.

Meanwhile, the shiny laser retro–reflector sat coyly in the powdery dust of the Moon's totally airless surface, providing a stark contrast between the brilliant technology of a sentient species and the death-like grayness of that world that seemingly forever orbits a paradise. Its surface resembled cremation ashes long forgotten in some dusty urn.

When Armstrong and Aldrin landed on July 20, 1969, they found big boulders blocking their assigned spot to settle down on. With only seconds of fuel left they luckily found a suitable place. But unfortunately their exact landing location was unknown to scientists on Earth. This meant the laser retro-reflector was lost.

When the order finally came on July 21st to try to locate the lost reflector, Rundle at Haleakela had the golden opportunity to get off the first laser shots at the Moon. Walker's laser at Lick was having the sabotage damage repaired.

Rundle commenced firing away searching for the reflector, but the answer kept coming back: "We don't see anything returning, Bill. Can you crank up the laser a little bit?" Turning up the voltage slightly,

Rundle winced, recalling his equipment's limited capability. Would the modifications he had implemented do the trick, or were they pushing their luck? These lasers operated at extremely high voltages. That and their enormous power were known to cause catastrophic failure. Still the answer came back: No luck. "Bill, we still can't see the reflector! Crank it up some more." Rundle had no choice but to comply and felt an uncomfortable sweat break out as he inched the power supply to even greater output.

Apparently the determination of the NASA crew got the better of them as they demanded maximum power.

Set Your Phasers to Kill

"Bill, give us everything you've got!" The NASA boys hollered as the Moon crept ever closer to the horizon. That night, the moon would only rise to ten degrees above the horizon, then quickly set. Therefore, time available was extremely limited. Unfortunately, NASA was just too worried about the Russians beating them to the Moon. Rundle's apprehension proved well founded as a loud POP! POP! POP! erupted from his equipment. He knew too well that this signaled the explosion of flash lamps and the possible breakage of ruby rods, a primary and very expensive component of the Korad Laser. As a smoky stench permeated the room, Rundle's hopes of making history as the first man to bounce a laser beam off the Moon's surface disintegrated along with his laser. Now of the three lasers originally available, only one remained operational.

We Have Company

Surprise, surprise! Astronomers could see other red laser spots appearing around the Apollo 11 landing site. Somebody else was trying to find the retro-reflector. It was assumed to be the Soviets, as no other country had the necessary technology. It certainly would not do for the Russians to successfully bounce a laser beam off the Moon using Apollo 11's retro-reflectors, stealing the American team's thunder, and it was no secret that the Soviets were following Apollo 11's progress with just this aim in mind. International pressure had been building over recent months, as the first country and/or organization to fire a laser at the Moon and get a reflection would realize a significant propaganda victory. The Russians needed such a victory to save face, having just lost the race to be first on the Moon.

They could then ridicule the Americans for having spent billions of dollars putting the laser reflector on the Moon while they, the Russians, captured the scientific glory.

The night of August 1, 1969, at Lick was absolutely gorgeous with crystal clear skies with no clouds to block the laser beam. Searching swath after swath for the target reflector was heating up Walker's laser. It had not been designed to run so long at full power and at maximum firing rate. Walker worried the laser would damage itself, as they were known for self-destruction. Laser powers in excess of a billion watts could blow a chip of ruby laser rod off the last amplifier surface or make a string of bubbles at the end of the rod. Both chips and bubbles were always in the center of the last ruby amplifier, so he'd know exactly where to look. The trouble was the laser was firing every thirty seconds and could not be stopped to allow an inspection for damage. The laser's power supply was overheating as well, so Walker hastily requested fans to blow air over the pulse-forming network, or PFN. The search continued, with one nail-biting pass after another.

Finally the laser beam struck the reflector, resulting in the first return signal in history. The historic date was recorded in a paper by Dr. Jim Faller, Irv Winer, et al. in the prestigious Science journal.[63]

Unlike contemporary NASA achievements, such as the Curiosity Rover on Mars, this one was not marked by high fives, champagne, or hugs. Everyone was exhausted and could think only of getting the heck out of there. But the event is commemorated by an exhibit at the Smithsonian Natural History Museum in Washington, DC. Walker is featured in this exhibit.

Walker, in charge of operating a Korad laser with his associates at Lick in California, thus received credit for being the first to range off the Moon. Simply firing a laser at the Moon was not significant in terms of doing science; that had been done in the early 1960s shortly after the invention of the laser. What was hard to do was to get the beam reflected back to Earth and to detect the reflection—the meaning of "ranging."

These days, after fifty years with no wind-driven dust or rain to mar the retro-reflectors, they are still perfectly functional and lasers

---

[63] "Laser Beam Directed at the Lunar Retro-Reflector Array," Faller, Winer, et al, *Science*, Oct 1969, vol 166, no. 390: 99.

are still firing at the Moon. The original Korad nanosecond ruby lasers have long since been supplanted by modern short-pulse pico- and femto-second[64] lasers, which produce pulses down to a couple hundred thousand times shorter duration than the ruby ones. This makes it possible to measure the Earth-Moon separation to accuracies of a *millimeter*, about the thickness of a paperclip.

Now Mother Nature's secrets can be even more deeply explored. Questions once beyond our ability to determine are now fair game. Does the Moon have a liquid core, or is it frozen solid as we have always assumed? Can we use this information to predict earthquakes? How much does the Earth wobble while spinning like a top on its axis? (Laser ranging has already determined that it wobbles a lot more than previously thought.)

Laser ranging has made it possible to determine with great accuracy the rate at which the Moon is gradually receding from the Earth. The action of plate tectonics has been verified as the relative positions of the laser-ranging observatories on Earth have been measured to be slowly drifting apart on Earth's crustal plates. The first successful ranging of the Moon by Korad's Hal Walker helped make these discoveries possible.

In the years since the first manned Moon landing on July 20, 1969, two more Apollo and two Russian unmanned missions left behind laser beam retro-reflectors. Scientists now have five to fire at. While Korad's Bill Rundle had to take second place to Walker when it came to winning the race to be first to get a return off the Moon, he still played a major role in the development of this technology, finishing the design and production of Korad's advanced moon-ranging McDonald laser. It was in action on a Mount Davis in West Texas for fifteen years from October 1969 when it first began receiving laser returns from the Apollo 11 reflector to 1975 before it was replaced. With the ongoing advances in laser technology taking place as we enter the second decade of the twenty-first century, it is thrilling to contemplate the discoveries that wait to be made, just over the horizon.

NASA had been late in placing an order for the McDonald laser, a model K-2600,[65] designed for lunar ranging. The lateness of the order

---

[64] A nanosecond is a billionth of a second, a picosecond is a trillionth, and a femto is a quadrillionth. It is said a femtosecond exposure would be fast enough to photograph an electron orbiting a nucleus.

was why it couldn't be utilized in the race to the Moon. As for the other two high powered lasers, the one at Lick was shipped back to the owner, Wesleyan University, and the Haleakala laser resumed spying on Russian satellites. The Spacerays laser was returned to NASA in Huntsville, Alabama where the Nazi rocket scientist, Werner von Braun, succeeded in getting his old enemy into space. .

With Fingers Crossed; Korad Scientists Surf the Wave

The year prior to the moon landing, 1968, had been a tight year for Korad, as management of the business had shifted in December 1967 from Theodore Maiman to Dr. Clayton Zerby, and this change was accompanied by the departure of virtually the entire engineering department. Dr. Jim Boyden, head of engineering, left with seven engineers to staff Holobeam Lasers in New Jersey. It was essential that Korad maintain its position at the forefront of research as a trusted subcontractor to the US government. Ben Parks, Korad's electronics whiz kid, was in the thick of the moon-ranging team on Mt. Haleakala and discussed the tenuous state of preparedness the team felt during that endeavor.

Certainly Korad was not the sort of corporation to let exhaustive preparations and careful deliberation cost it the opportunity to be involved in such headline-grabbing work. It would not do to present itself to NASA and the rest of the scientific establishment as anything other than a mature and fully prepared operation, lest the contract go to a competitor.

Parks related, "We had to deliver the Haleakala laser so the company could meet its 1969 shipping quota. It was shipped early to

---

[65] Rundle's three-nanosecond K-2600 laser-ranging system was described in his technical paper "Performance of a 2GW Ruby Laser Designed for Lunar Ranging" presented at the EIA-US Dept. of Commerce, Paris Laser Colloquium, Nov. 18, 1969. It was a souped-up, faster laser than its little brother, the K-1500. It fired a pulse every 3 seconds (instead of 30) and had a beam divergence of 1.2 milliradians, quite important for getting a return reflection because it produced a laser spot 1.5 miles wide (2.4km) on the Moon. It had three laser heads, an oscillator with a 100mm X 10mm ruby rod and two power amplifiers each with 200mm X 20mm rubies whereas the K-1500 had only one amplifier laser head. The K-2600's 3 nanosecond pulse was over three times shorter than the K-1500's 10 nanoseconds, the industry standard. The shorter pulse gave three times better accuracy of the lunar distance, to15 cm (6 inches). The K-2600 sported 1.8 billion watts with 5.5 joules of energy in each pulse.

make the books look good for that year. Everybody knew it wasn't ready to go."

Everybody, it seemed, except the customer. Parks continues: "Everybody in the factory knew it was months away from being ready to ship, but we shipped it. Then we went to Hawaii to make it work. We had to work on the optical and electronics designs and many other things. There was a lot of shipping of parts back and forth. I went back and forth a lot on that red-eye midnight flight many times."

When Parks and the Korad crew were perched up on Mt. Haleakala, they were laser illuminating Soviet Russian satellites for the purpose of photography. The laser was doing the job of a photographer's flash bulb. Parks went on to say, "And we even read the serial number off of it." But this was mild compared to what the boys did *to* a Soviet satellite: they blinded it permanently with their high-power ruby laser. Nothing was said or written about this in the media. The response was a replacement satellite that the Haleakala boys again tried to zap, but it kept operating despite numerous attempts to blind it too. Obviously the eyes of the Russian satellite had been equipped with laser safety glasses to block the beam.

Union Carbide's physicist and at that time president of the Electronics Division (including the Korad Department), Dr. Bob Charpie, saw an opportunity to get into military lasers. The Vietnam War was raging and Charpie understood both laser guided smart bombs, also known as target designators, and laser radars. He approved forming a military group at Korad. However, we executives at Korad well knew the government's intentions were to use Korad to quickly develop laser weapons, then have them built by aerospace giants like Hughes in nearby Southern California and Martin Marietta in Florida. How did we know that? Simple. Uncle Sam had insisted on a clause in the contract forcing us to supply them with a complete set of blueprints.

Next is a war story about the adventures of Carl Schulthess. Korad contributed to the first use in warfare of pinpoint-accuracy laser target designating weapons. This technology helped the United States win resounding victories in the Gulf Wars. Back in 1972, an article in *Laser Focus* magazine said that smart laser-guided bombs provided an order of magnitude (ten times) improvement in accuracy over the "dumb" weapons previously used.[66] The article also stated that the Pentagon

---

[66] "Those Smart Bombs," *Laser Focus* (July 1972): 6.

was exulting over spectacular successes with laser-guided bombing in Vietnam.

Chapter Seven: The Carl Schulthess War Story

❦

Carl Schulthess

After Maiman invented the first laser at Hughes Research Labs (HRL) in Malibu, the environment there changed to one less hospitable to pure research and development. While this focus is understandable, it did tend to cost Hughes the services of a number of its core players in the development of the laser, Maiman being the most glaring example. Korad sprang from this bloodletting as Maiman lured away a few of Hughes' talented people, and a plethora of commercial applications followed. None of them, however, kept Korad's crew as busy as the lengthy wish list of its primary client, Uncle Sam.

The Pentagon had high interest in this new laser technology that held the promise of revolutionizing the way its military targets were designated, the way bombs and missiles were guided to those targets, and the way laser radar supplied the range to a known target simply because the tiny laser spot was placed directly on the target. A crisp laser beam targeting an enemy provided a pinpoint guide for a flying bomb and was galaxies away from how this process had been undertaken in the past. The difference between the point, shoot, and pray method of the twentieth century's earlier conflicts with its haphazard hit or miss result, and the lethally decisive performance of today's "smart" bombs is due to the tailor-made introduction of the laser.

These smaller bombs were not only lethal and smart, but their more accurate targeting produced less collateral damage with fewer innocent bystanders killed. Beginning in 1967 these weapons were used in combat with laser-guided missiles[67] to target the increasing effectiveness of North Vietnamese antiaircraft guns. When we use the term *point and pray,* one might think we are exaggerating. Surely the techniques for designating or illuminating targets and guiding munitions to them were reasonably sophisticated prior to the introduction of the laser. One would be wrong, though, as the following story about the first night time use of the smart bomb in Southeast Asia illustrates.

Korad Lasers Light Up the Ho Chi Minh Trail in 1970

Even though he could not boast of a college degree, Carl Schulthess[68] was one of Korad's most talented technicians who had managed, through a combination of natural talent, dogged determination and a pinch of good luck, to land a good job in the growing field of laser applications. Perhaps part of his success derived from his commanding presence; he was a big, good-looking man with a crew cut. His speech itself was remarkable for its rapid-fire delivery and authoritative tone. Not one to dwell on personal issues, he was all tech and straight to the point. Still, Korad's latest request, made in the winter of 1969, evoked some excitement. They wanted him to take a bunch of Korad's KY10-A lasers to an American base in Ubon, Thailand, so that they might be directed at targets from aircraft.

---

[67] *Encyclopedia of the Vietnam War,* ed. Stanley Kutler (New York: Charles Scribner's Sons, 1996) 498.

[68] From an interview with Carl Schulthess on May 17, 2009.

The lasers would illuminate the targets along the Ho Chi Minh Trail, the Vietcong's primary route of resupply during the Vietnam War. Then rocket propelled missiles and flying glider bombs would home in on the reflected tiny laser spot for super accurate targeting never before attained.

The American military already had daytime laser illuminators,[69] which didn't work at night. So the enemy had simply just switched to night operations, and many supplies were getting into South Vietnam at night. Gaining nighttime capability was crucial to stopping the flood of weapons pouring down. The lasers were scheduled for shipment in late December, and Schulthess was given only a week to get all his shots, his passport, and his affairs in order. Schulthess recalled getting every shot under the sun. "I just about passed out after a couple of days. I was pretty sick."

If the process worked as the Pentagon planned, F-4 fighters could use the laser beam to guide their munitions to the target. But Thailand—wow!—this could be an adventure beyond babysitting a couple laser devices: a tropical climate, little personal supervision, and slender Thai ladies.

Malcolm (Mal) Stitch, 1973

[69] Daytime target designators were mounted mostly in little single-engine Piper Cub type airplanes carrying a forward observer. They were expendable, as Dr. Art Guenther, in charge of lasers at Kirtland AFB in New Mexico, told me—$10,000 for the plane and the same for the pilot.

A Bachelor's Adventure

Carl Schulthess recalled some specific instructions from Korad vice president Dr. Malcolm Stitch in our interview on May 17, 2009. Initially Korad's prize supertech,[70] Jim Rose, had been selected to head up this assignment. "We were to fly to a Royal Thai Air Force base, which was one of several military bases along the Mekong River that separated Laos from Vietnam. It was near the three-border meeting of Laos, Vietnam, and Cambodia. (The Vietnam border was about 150 miles directly east of Ubon.) As the river went farther north, it was Thailand on the western side of the peninsula and Laos on the eastern side; the river separated the two countries. We were to babysit and install the KY10-A at Ubon Airfield in Thailand."

Alas, Rose's wife had considerable reservations about her husband spending a month or two in a war zone, so he was taken off the project. Stitch's approach to Schulthess involved asking him if he was a single man. "Would you like to go on a trip?" He had no hesitation and promptly accepted without having even dipped his toe into the water. So now here he was, about to go halfway around the world to babysit three or four KY10-A's as part of a military operation, and he was not even part of Korad's military group; what's more, he had never seen a KY-10. But fearless fellow that he was, off he went!

The first leg of the journey took Schulthess from California to Eglin Air Force Base in Florida, where he met with Rose, who ran him through the KY-10 schematic drawings for two days as they took in the area beaches. The laser, about the size of a large shoebox, was to be mounted to the top of the Starlight Scope, itself the size of a man's leg. The scope, even by itself, was too large to fit into the cockpit of the F-4, hence its use in the C-130 gunship that sported four turboprop engines.

Looking over the drawings, it seemed to Schulthess as though the setup might be somewhat unwieldy; the infrared scope with its attached laser was to be mounted on a tripod and gimbal with the operator aiming the entire contraption while gripping two handlebars attached at the base of the scope. While it seemed to him that such a system may prove to be exhausting for the operator, with so much

[70] Supertechs, talented junior engineers, are non-degreed jacks of all trades. Korad had quite a few, such as Ray Larson, Dennis Mulvaney, and Steve Berry. They are rare and quite valuable to their employers.

weight to be managed from the handlebars. It still had a certain *Easy Rider* movie quality to it; he could almost imagine Dennis Hopper or Jack Nicholson circling the Ho Chi Minh Trail from 10,000 feet, astride their trusty target designator as the wind whistled through the C-130's open side door and through the intrepid operator's hair.

Jim Rose slid his tongue around the rim of his final margarita, the grains of salt crunching between his molars as he pushed his chair back and bid Schulthess good night. Contemplating his first flight on a C-130 slated for the next morning, it seemed to Schulthess that between his talks with Rose and Korad's Don Smart, who was part of the military division and regarded as quite sharp, he was as comfortable as possible with his assignment. Now for a good night's sleep.

Cavemen in Danger of Extinction

Korad's KY10-A wasn't the first laser employed in Vietnam; it was just the first to offer an effective nighttime capability. The F-4s delivering their deadly payload along the Ho Chi Minh supply route had a compact, daytime-only, target-designating system consisting of a visual-only telescope bore-sighted to a ten pulse-per-second Martin Marrietta, Florida laser. The planes flew in tandem and laser designated for each other. These pilots got so good at the process that when the Vietcong antiaircraft guns rolled out of their hillside caves on railroad tracks only to dart back in for protection, the F-4s back-seater placed his designator spot directly in front of the mouth of the cave. Then, as the bomb approached the ground, they walked the beam toward the cave entrance. Then the bomb's steering vanes flattened out its steep dive into a more horizontal flight and directly into the depths of the enemy's hideout.

Blind Bats

As effective as this system was, it was not designed for nighttime use. The only nighttime operation involved the use of the C-130 transport aircraft, which had been fitted with the infrared Starlight Scope, useful in spotting the heat signature of the enemy vehicles but worthless for guiding munitions to their targets. The Air Force's solution to the problem of target designation at night was comical in its simplicity: when an enemy vehicle was spotted in the Starlight Scope, which jutted awkwardly out of the right side door of the C-130, the crew dropped three flares off the rear loading ramp, which remained open during the low-speed flight of the C-130. The crew could determine the exact position of the target from the flares which were visible in the inky blackness, then radio that position to the F-4 pilots. The problem was the resourceful VC quickly became aware of the technique and simply changed their direction or stopped moving entirely, which obviously made finding their accurate location all but impossible. The hit percentage was estimated as dismal as single digits, hence the name given to the C-130 squadron: *The Blind Bats*.

In addition to the pathetic performance of the flare system, the Air Force was concerned about the danger posed by the existence of so many flares on board the C-130. Should one come under attack, the potential for a raging fire and subsequent crash was unacceptable. The C-130 crews had a contingency plan that itself bordered on the comical: In the event of incoming rounds or an already existing onboard fire, they would point the craft's nose skyward and climb steeply while pulling a rigged release that secured the crated flares in place in the craft's cargo bay. At this point, gravity would assist the crates in exiting the back loading ramp. Hardly the type of arrangement one would refer to as high-tech, but hey! It worked.

To improve upon this Rube Goldberg arrangement, the Air Force and Autonetics Division of the Rockwell Corporation[71] in Anaheim collaborated on a project to attach a laser to the top of the Starlight Scope. This should, they reasoned, be far more effective than dropping a crate of lit flares from 10,000 feet. Autonetics subcontracted with Korad to provide the KY10-A laser. It would blow the whistle on the target, illuminating it for the benefit of the lurking F-4, which could then release the flying bomb and zap the target within a thirty-foot "probability of error" circle. The result was over 80 percent direct hits, with the majority of the 20 percent failures being caused by "dud" bombs.

An Elite Sapper Squad Strikes

The next morning Ubon reminded Schulthess of an anthill, as a multitude of Thai workers covered every square foot of the base, performing any number of duties from food preparation to runway maintenance. Schulthess was to work with Clyde Board, Autonetic's

---

[71] Now Boeing

man on the project. It took both men a couple days to get themselves and their equipment organized and to get the lasers bore-sighted to the infrared scopes. Such a pairing of devices had hitherto not been attempted and posed a significant challenge. Moreover, the VC were being uncooperative as well.

The second day after they had the systems in the planes and the first couple missions had been flown, Schulthess arrived at work and was told there had been a Vietcong sapper attack on Ubon involving twenty men carrying explosives-laden backpacks. Called "*đặc công*" ("special task") by the Vietnamese, sappers were a force economy measure that could deliver a stinging blow. They were a series of elite groups, especially adept at infiltrating and attacking airfields. The VC had employed sapper squads to hit hard and fast and get out quick. Their heavy losses during the Tet Offensive had made large-scale attacks hazardous.

In this instance they targeted the C-130s specifically, while ignoring the F-4s and the munitions and fuel supplies. There was no question at that point but that the Vietcong were well aware of the laser program and its whereabouts. Terrific, Schulthess thought—now he could add the potential for armed attack to the considerations keeping him up at night. *Next thing is they come for me and there is a $5,000 bounty on us civilian 'Tech Reps*, he thought. He had been told to travel in groups and to stay sober and alert.

Everyone was charging about, nobody seeming to know what was going on as Schulthess and Board hurried out to the airplanes to inspect the equipment and see if there was any damage. C-130s were used for a variety of things, mainly cargo and gunships. In an old pickup supplied to them for getting around the base, the two men motored over to their C-130. The planes were intact and unharmed as the Thai and American police guarding the base had been tipped off by a local farmer who saw the attackers sneaking by at 4:00 a.m. It seems the fellow had gotten up in the morning to take his water buffalo out to graze and he saw a pack of men walking through the fields. He had run over to his village police, who contacted their counterparts at Ubon.

The base police were waiting for the VC sappers when they got through the fence. If not for the timely response of this local farmer, the attack might well have been successful. Even with the nasty reception the VC received as they were breaching the dog-patrolled perimeter fence where the bulk of the intruders were shot, a few were still able to reach the vicinity of the C-130s. Their bloody, bullet-riddled corpses provided evidence of their zeal for the benefit of Schulthess and Board's unaccustomed eyes. "They were badly shot up." Schulthess recalled. "They had big knapsacks on their backs loaded with explosives. All twenty were killed."

The balance of the day's work consisted of filling and stacking sand bags to serve as a breastwork for protection of the C-130s. The planes flew again that night. This sapper attack came about two weeks before the Vietnamese holy day, Tet, which motivated the Air Force to fly the entire squadron of a dozen or more planes to a safer airbase in Bangkok.

On Schulthess' initial flight from the States, the C-130 crew had picked up an Air Force major who was to serve as the team's liaison to the military. One night during dinner at Clark Air Force Base, the major asked Schulthess about his background. Proud of his work, he promptly mentioned the laser, which drew a hasty remonstration from the spit-and-polish officer, who advised him in no uncertain terms to *NEVER* use that term in public. Schulthess had thought the Major was overreacting, but now with the passage of time his reaction made perfect sense.

Congratulations! Now Here Are Your Crosshairs

Schulthess and Board worked with the aircrews in calibrating and fine-tuning their new target-designating creation and were realizing considerable success. While several squadrons of the Blind Bats continued their nighttime operations using the Mickey Mouse flare system, another squadron employing the Korad KY10-A laser mounted on the Starlight Scope setup that Schulthess and Board had built was nailing the VC at a hit percentage of 80-plus percent. Because the laser was invisible and the C-130 flew at 10,000 feet, the Vietcong had no clue what was about to ruin their evening.

With the new setup, the C-130 circled the target[72] after first acquiring its heat signature through the Starlight Scope. As it continued to circle the target, its path described a cone, with the tip of the cone being the target. The laser was turned on and then the

[72] Tight circling was the C-130's claim to glory. It mimicked a helicopter's ability to loiter over a target but as it flew much faster, it got to the target quicker.

accompanying F-4 was told to drop their bomb or missile, which had a laser sensor on its nose. This allowed the weapon to home in on the laser spot held on the target as it moved. The sensor on the front of the bomb was divided up into four quadrants. The intensity on the upper two quadrants was compared to the lower two quadrants, and likewise for the left two and the right two. The bomb had fins on its tail that moved automatically to null out the differences in intensity between the two sets of either the vertical or the horizontal quadrants. The bomb was steered, so as it was dropping, it didn't matter where the trucks went as long as the laser beam was illuminating the lead truck, or whatever truck they wanted to hit. The bomb steered itself directly at the reflected beam illuminating (or designating) that truck. As Schulthess said, "These guys actually got so good at it, that the VC installed radar-controlled antiaircraft guns along the trail. The F-4s would get shot at every once in a while by their radar-guided guns."

Please Remain Seated

Deploying and testing the new capability was not without its hair-raising moments. One particular operator had seated himself behind the designator's handlebars after ignoring explicit instructions to wear only a fanny pack parachute that was less likely to hang up on the laser designator's various parts as the entire apparatus was sticking out the aircraft's side door. This individual had decided he would be more comfortable wearing a front-pack chute. Unfortunately, as the C-130 was cruising along 10,000 feet above a pitch-black jungle infested with a ruthless enemy, the D-ring on his chute became snagged on one of those Easy Rider handlebars. The parachute deployed out the side door of the aircraft. After a moment or two of ripping, clutching and thumping, its wearer followed it. Fortunately for him, he was equipped with a locating beacon that enabled his retrieval the following day. It was speculated by his fellow airman that his lonely night in the jungle may well have made him more favorably disposed to following orders for the balance of his tour.

The success of the new laser met with considerable enthusiasm among the aircrews and the brass; the latter had been the attentive recipients of Schulthess' periodic dog-and-pony tours, during which he and Clyde Board explained the status of the system to generals who were trying to determine the feasibility of the new device. Korad's KY10-A performed wonderfully; indeed, the whole system worked very well, and both men received considerable credit and praise, not simply for supporting the new designator, but for their efforts at selling the system afterward.

From Hero to Smuggler

Excited about the program from the beginning, the aircrews considered the team to be heroes for getting the direct-hit percentage to 80 percent. This success led to a new contract for Korad to supply lasers that were to be mounted on a more stable and targetable platform.

The men remained busy supervising and maintaining the target-designating system as flight after flight took off and returned. While the crew of these aircraft were courting some risk, as our hapless parachutist mentioned earlier attests, they still were somewhat insulated from the killing as compared with the Army soldiers on the ground who did their killing eyeball to eyeball as opposed to up in the sky. Because of this insulation, as well as the fact that the Air Force was a volunteer service—those drafted went into the Army—the war was virtually a nonissue during lunch breaks. Most of the Air Force personnel seemed not to share the rabid opposition to the war typical of stateside protesters, preferring a more pragmatic approach. While they were as appalled as anyone by the body count, they felt that there were some reasonable principles involved and that to uphold them, a country must occasionally fight a war, like it or not.

All Hands on Deck!

The proceedings at the Ubon Officer's Club may have been standard operating procedure for military personnel, but for Schulthess, a civilian, they were anything but predictable. One of the games frequently played at these gatherings was called Carrier Landings, where they lined up twenty to thirty feet of tables and proceeded to pour beer over them all. The squadron then would pick up the soon-to-be-civilian, swing him by the arms and legs, and hurl him down the length of the tables. The object of the game was to remain stable and centered on the tables, and the drunker they got, the more frequently they would careen off the side of a table and crash land on the floor. Schulthess thought it an altogether appropriate sendoff.

The collaboration on this target-designating system between the Air Force, Autonetics, and Korad was a stunning success, and even if it did not completely stop the flow of arms from North down to South

Vietnam, it certainly made the job of smuggling them south far more difficult. The system's success also instigated research and development of the next generation of "smart" technology that followed the Vietnam conflict. Someone once described the Ho Chi Minh Trail as Broadway at night, with its constant stream of headlights. Schulthess' team had the satisfaction of knowing that they were able to extinguish the glow along their section. The VC apparently considered the system worrisome and devoted a highly trained sapper team to its attempted destruction. Certainly not all similar designation systems fared as well.

A competing defense company had been contracted to deploy a low-flying craft bearing a very sensitive antenna. The craft was to be flown parallel to the trail, detect the electromagnetic flux of a truck's ignition coil, and talk the F-4 pilot in for the kill, similar to the process followed by the Blind Bats. As Schulthess learned one night as he listened to the audio of one very nervous pilot as he attempted to return to Ubon in his shot-up plane, the low altitude proved to be a costly requirement. The pilot was advised by the tower controller that heavy traffic required that he circle again before landing. After a long string of serious swearing, in which the pilot responded he would land on top of an F-4 if necessary; he brought the plane in for a safe if not smooth landing. Viewing the plane the next day, Schulthess noticed that all of its rear control surfaces were inoperable, having been shot full of holes. The pilot had steered the plane with only the throttle and the flaps, commonly referred to as the "incentive plan."

In casual talks with the higher brass, Schulthess at one point suggested that the Department of Defense consider putting a bounty on destroying the enemy's assets, thus permitting the free enterprise system to work its magic. Various defense companies could make investments and bring the best hardware they could to the action. They could then pay people a competitive wage to operate in the hazardous war environment and get a good return on their investment at the same time. The Air Force could be employed to take photos of the battleground for the purpose of accounting, he reasoned. The brass responded that the DOD had strategic concerns with using some of the latest and greatest technology because they did not want the Soviet Union to pick up any clues as to our advanced capabilities.

After ten weeks amid the jungles and villages of Thailand, Schulthess was glad to return home to Santa Monica and his less hair-raising duties at Korad. After all, he did have his new Nikon to play with. What more could a young bachelor ask for? Schulthess did not remain single for long, however, as he recalled in our recent interview.

"I was dating somebody when I left, and I was uncertain as to whether I wanted to get married or not. I had been dating her for a couple of years and when I was over there, I realized that it was a lot more fun being with her than it was being alone. So when I came back from Thailand, I came back with a ring and proposed, and that was forty-two years ago, and she's sitting over here," he added, wrapping an arm around his spouse, Jacque.

Additional Military Applications; Some Purely Defensive

Aircraft were not the only vehicles equipped with these lethal new components, as Hal Walker, now retired from an illustrious high management career at Hughes in military lasers, has pointed out. There was also the Abrams tank. "Lasers allowed us to have what was called 'Force Multipliers,' meaning that electro-optical and laser-based fire control systems gave our side an advantage that we called Force Multiplication. This is where the other side had more guns and more tanks than we had, but they didn't have the same accuracy of shooting. In weapons platforms like the Abrams tank we had a first-hit capability and high probability of destroying the target. And this was because we were using lasers." According to Walker, that was one of the technological breakthroughs that broke the back of the Soviets in the arms race. They could not keep up with the innovations emanating from American industry and eventually threw in the towel, signaling the end of the Cold War.[73] Hughes for a time was the world's major producer of laser systems, making them by the tens of thousands. We had proven the laser's effectiveness in the Vietnam War, thus ending the Cold War with its potential to kill hundreds of millions of people. Perhaps those killed in the Vietnam struggle did not die entirely in vain.

Carbide Sells Korad's Military Division

With the ending of the Vietnam conflict, military business tanked. So at the end of Union Carbide's ownership, Korad laid off about fifty of their

---

[73] It wasn't just these low-power early military lasers that helped end the Cold War, it was Soviet Russia's fear of "Star Wars," Reagan's touted plan to shoot down Soviet ballistic missiles with high-power lasers in big airplanes like the Boeing 747.

military people. By the end of 1971,[74] Carbide had sold the military division to North American Rockwell (now Boeing).[75]

As worldwide research on ruby lasers died out, next to go were the scientific lasers. The laser business moved on to more efficient and powerful lasers, such as YAG, diode, optical fiber, excimers, pico- and femto-second. Gradually over the years, Korad was to concentrate on industrial laser machines, mainly welders, drillers and markers.

However, ruby lasers did not completely disappear, they live on to this day in such applications as tattoo removal, double pulse holography and diamond drilling. One can still buy such laser systems, check the Internet to verify.

All Work and No Play?

In light of all these sophisticated developments and amazing capabilities, a person might envision the average industry lab as being staffed by stuffy and eccentric eggheads, prone to lengthy and esoteric discussions of technical specifics. In many cases that image was fairly close to the mark. But the old expression "boys will be boys" applies to technicians and PhDs as well as to the rest of us, and the personnel at Korad Lasers were no exception. In fact, at times it seemed as if their particular choices in mischief were as ingenious as their achievements in science, as the following chapter illustrates.

---

[74] This according to Jim Scaroni, who joined Korad Jan 2, 1972.

[75] "Rockwell Taking Korad Arms Line," *Laser Focus* (May 1972): 4.

Chapter Eight: Engineers Gone Wild

Clifford B. (Cliff) Cordy Jr.

Cliff Cordy, who has a doctorate in electronic engineering,[76] has extensively edited and contributed to this chapter. I thank him.

However, when at Korad he was not yet Dr. Cordy. The photo below offers the proverbial thousand words' worth of Cliff, as we used to call him. He is wearing a Peruvian Alpaca wool headdress essential for keeping your head warm up in the Andes Mountains where the air is both thin and cold. Thin air, you might know (I didn't

[76] Cliff Cordy's educational background is highly unusual; he holds three bachelor's (BS) degrees in math, physics, and electronics engineering, having earned all in 1960; a master's (MSEE) '63; and a doctorate (PhD) '95, all from Oregon State Univ.

until Cliff informed me), slows your metabolism, causing you to get cold.

Various Koraders Nurture Their Inner Child

What does a seasoned laser engineer (or his supervisor) who typically had a master's or doctorate in physics do on those occasions when he needs a break from his sometimes tedious and taxing work? Let us undertake to answer that question by first considering the case of Korad's prized supertech (not a college-degreed engineer), Jim Rose, and his favorite hot rod. The reader may recall that the decade of the 1960s took place smack in the middle of America's love affair with the automobile. Rose was particularly proud of his chosen rod: a mid-1950s Volkswagen Beetle that was already old when he sported it around Santa Monica in 1964.

Even though it boasted (or perhaps whispered) of a hefty 36 horsepower, Rose's bug was fast off the block as it had a powerful first gear and had barely any mass to it. When responding to a stoplight challenge against many a larger and heavier vehicle, he was able to use these characteristics to his advantage. His Volks would hop forward like a bug while his antagonist burned rubber for a moment as it struggled to overcome its inertia. However, he was a special type of hot rodder. He liked to sneak up to the right of people—in the parking lane—then lurch out in front of them when the light turned green. He was convinced a VW bug was the fastest car ever built, but only from zero to a typical human's running speed, which was as fast as those old VWs would go in first gear. Nobody he beat was ever burning rubber. They just weren't paying much attention to the stoplight.

Cordy used to delight in challenging Rose's perceived invincibility. "The only reason you can beat a race car like a Chevy Corvette is because he isn't trying," Cordy goaded.

"No, no, I go fast!" Rose insisted, and on it went, back and forth for over a month until Cordy showed up at Korad behind the wheel of a brand new Corvette. As in *High Noon* at the OK Corral, the standoff was inevitable. One night after work found Rose and Cordy rolling to a stop alongside one another at a stoplight near Korad. Rose, convinced of his calculations, cast his intrepid adversary a smug glance as Cordy let loose a throaty roar from his 300-horsepower monster. As the Volkswagen sounded more like a sewing machine when revved rather than an African lion, Rose thought better of issuing a reply. Instead, he put all his faith in a quiet, unassuming display of confidence. Recognizing his strength in this battle had always been in the short run—the extremely short run—Rose issued his challenge: a drag race to the other side of the crosswalk. It's only ten feet, he reasoned. No car is going to beat my bug's acceleration in a hopping contest.

As for Cordy, it certainly would not do for him to be bested by this antique rattle-trap, yet he had just picked up his Vette and had hardly mastered its high gear ratios and light flywheel. What if he revved it and dumped the clutch only to have it die on the spot, leaving him the laughingstock of all bystanders? Clearly only one strategy was OK: floor the baby as soon as the light changed and then immediately dump the clutch. A quick prayer for the protection of his male ego might also be in order. This determination made, the light changed and Cordy's Vette roared from the intersection like an angry F-18 Hornet off the flight deck of a carrier, leaving a stunned Rose wide-eyed in his wake.

"I watched Rose cross the crosswalk in my rearview mirror. That is *not* an exaggeration," Cordy noted afterward. "And I think he would verify that." When Rose sputtered in disbelief and expressed admiration for the Vette's performance, Cordy calmly advised him about the finer points of independent rear suspension, which the VW had too, but without sufficient power (only 36 HP) to make any difference. When Chevrolet put the independent rear suspension in their 1963 Corvette, he consoled his friend, the wheels no longer just sat there and spun.

My Vette can Whip Your Vette

Cordy's superiority in the Crosswalk Hop was soon threatened by the arrival of new hire Steve Berry,[77] who at age twenty-four had just come out of the U.S. Navy's submarine service as an electronics tech. He proudly drove a beautiful yellow '65 Vette with a 360-horsepower engine, as compared to the mere 300 in Cordy's Vette. Cordy escaped the inevitable showdown when Berry's '65 was promptly stolen. Not to

[77] While working at Korad, Berry earned a bachelor's degree in engineering. With the collapse of the military at the end of the Vietnam War, he lost his Korad job in 1972. Then he became a sales engineer, where he did well financially.

be thwarted, Berry replaced his 360-horse baby with an even more magnificent Vette, a jet-black beast that packed five hundred.

Cordy was occasionally asked to accompany Berry on cruises around Los Angeles. One evening, as they walked out of Korad, Berry invited his friend to drive his beast. Cordy had been waiting and hoping for this opportunity, so with no hesitation, he took the wheel and found 500 horses to be altogether more than he was used to. "My Vette was impressive," Cordy later noted. "This thing was *awesome*." Clearly this was the Starship Enterprise on steroids.

They came to a stoplight just before the Santa Monica Freeway, the end of the I-10 running from Florida to California. Cordy noticed that if he went directly straight across the street, he would be on the onramp, and this would be a good place to get a feel for the acceleration of this monster. Not wanting to risk any mishap with his good friend Berry's pride and joy, he simply engaged the clutch, the same as he had been doing, then he nailed the gas pedal. While he was carefully letting up on the clutch pedal, a nothing special Chrysler Dodge charged by into the intersection, obviously bent on beating the two friends onto the onramp. Already planning an acceleration test, Cordy put his high rpm's in action and took off like a rocket by immediately releasing the clutch.

As he relates it, "I had planned what I would do, and I did it. When I nailed the gas pedal, the driver of the Dodge must have felt like Jim Rose in his VW, but this time, that was not my plan. All I did was step on the gas pedal. Steve had done that a couple times while I was riding with him on previous occasions, but I had no clue what a dramatic effect it would have on me when I was behind the wheel and had to control the process."

When Cordy mashed the gas pedal to the floor, he experienced what he subsequently referred to as tunnel vision. "I was literally hanging onto the steering wheel. We were three-quarters of the way across the street as the tach sailed past six thousand rpm, and we were going fifty miles an hour." The Dodge was somewhere totally behind the two men as Cordy dropped the beast-Vette into fourth and throttled onto the freeway. This was clearly way past something he could handle on short notice in his car, he mused to himself. After a couple more off- and onramps, relishing the capability of Berry's new toy, Cordy was persuaded to return to the Korad facility after nearly mowing down a hapless Volkswagen that had escaped his attention as he flashed past at warp twelve.

Less robust vehicles in the vicinity were granted a reprieve when, like its predecessor, Berry's new black Vette was stolen—the all-too-usual fate of Corvettes in Los Angeles. And then, to top things off, just a year later Cordy's Vette was also stolen, but he got it back undamaged and drove it another decade.

Cordy sums up thusly: "I never raced a car in my life, except for that race with Rose for the width of the crosswalk. As far as I know, he never raced anyone either. It is just that when you have a car that will actually move, it is fun to move it occasionally. And Rose in his VW lurching out in front of an inattentive grandmother in her Buick is hardly a capital crime, even if it really isn't legal."

Cat Feeders and Anti-Vettes

Cordy's friend and Korad technician Ray Larson, a slender and unassuming former school teacher, provides an auto story in stark contrast with the foregoing tale. Larson could do just about anything, and generally it was innovative. He had a thorough grasp of a wide range of knowledge and was adept at applying it. While at Korad, Larson built an electric car, dubbed by some as "Raymond's anti-Corvette." Cordy found the contraption a source of ongoing amazement as well as consternation. *Why did Larson not make the vehicle small?* he wondered. Larson purposely designed it to look like a 1910 electric with high windows and a ton of air drag. It crept along at a top speed of thirty miles an hour with a tailwind. Quoting his friend, Cordy said, "If you're driving something that doesn't go very fast, it should *look* like it doesn't go very fast, so other drivers know what to expect."

Built on a Crosley frame with the original suspension and brakes and loaded with a pile of lead batteries and constructed of polyester resin and burlap bags (as this material was easier to tool and work with than the more typically used fiberglass), Larson's creation was as aesthetically appealing as Frankenstein's monster. Naturally, the heavy batteries challenged the brakes. Reasoning that such a creature would require an imposing voice, Larson mounted the loudest electric truck horns he could find. He eventually rolled it from the shop and into the California sunshine where the consensus was, as he hummed around the block, "It's *alive!*"

Larson told Cordy, "When I honk, people get out of the way. Only after they feel safe do they take time to look for the truck coming at them. In an emergency, it was important that people get out of the way, hence the horn."

Larson was a bit eccentric and given to creative pastimes, as was demonstrated by his subsequent innovation, the Larson Cat Feeder. He owned a friendly cat named Sylvester, whom Larson endeavored to keep supplied with tasty songbirds. In the front yard Larson installed a bird feeder consisting of a slender wooden post about fifty centimeters tall, topped with a small, circular wooden table loaded with bird seed. Unfortunately, Larson's feline had considerable difficulty landing a bird from his disadvantaged position on the ground underneath. This necessitated further innovation on the part of Larson, the intrepid owner.

At dinnertime, when kitty sauntered over to the cat feeder and lay down under it, the birds paid hardly any attention because they were used to their furry interloper. Larson experienced a moment of inspiration and retrieved his pellet rifle. He slowly slid open the glass door that looked out on his front lawn. *Ping!* Tweety fell off the edge of the feeder and into Sylvester's obliging mouth.

As it happened, Larson had a nearby grandmotherly neighbor who was a bird lover. Occasionally he spotted her coming along the sidewalk and would hurry kitty out the door. With a Pavlovian predictability, Sylvester took her usual place beneath the Larson Feeder just in time for Granny's arrival. *Ping,* and Granny's jaw dropped as she rushed to Tweety's aid, who was experiencing the indignity of a beheading at the hands of Larson's beloved cat. While his relationship with this particular neighbor suffered long-term impairment as a result of his most recent innovation, Larson's pampered feline never again missed a bird treat.

Cigar Wars

Another incident demonstrating technician Larson's resourcefulness involved his response to the encroachment of several of Korad's cigarette-smoking staff into what he regarded as his sacred, smoke-free domain. The engineers in Korad's military applications group were located in two rooms at the west end of the company's production area. One room was for those who smoked, the other for those who didn't. The two groups were separated by a door which Cordy took upon his own authority (which was nil) to close, using the excuse that his nonsmoking room served as a clean room for optics assembly. To ensure the door remained closed, Cordy removed the doorknob, leaving +a hole in the door.

Next to this door, Ray Larson assembled Korad's Pockels cells,[78] a very valuable commodity, an unpatented Korad invention, and a most unique Q-switch, making Korad lasers the most powerful in the world. Because fastidiousness was one of Larson's eccentricities, he carefully maintained his spotless workbench. The doorknob hole provided just the opportunity for mischief, which the engineers and technicians on both sides subsequently found irresistible. The hopelessly addicted techs, overcome by a desire to tease Larson and the nonsmokers, would blow smoke through the hole.

The stamina of the smoking staff was remarkable as they continued to exhale long drags of acrid tobacco smoke through the knob hole for the torment of our heroic Larson. Not to be outflanked at this point in the battle, Larson let fly a blast of his canned air spray through the hole and into the adjacent lab.

The coughing and hacking had barely subsided when the smokers conceived of an even bolder offensive. Putting flame to a large cigar end-to-end, they shot the flaming cigar through the knob hole for the further benefit of their intended victim. Larson retaliated by grabbing a pocket knife from its appointed place among his tools. Stabbing the side of his aerosol weapon, he launched the can back though the hole into the midst of his tormenters, who scattered in every direction as the deodorant container hissed and spun about their lab. After this escalation of hostilities, Cordy, as engineer-in-charge of the techs, told them, "These smoking games and wars are going to stop immediately." And stop they did.

Dr. Fred Burns, big boss #2 and operations manager, was fond of practicing Management by Walking Around (MBWA), many times smoking a big stinking cigar. This was in stark contrast to Maiman who stayed in his office. Once, Ray Larson followed Burns through the clean

---

[78] The Pockels Q-switches had to be kept clean. Located between the two end mirrors comprising the laser cavity, any dirt, fingerprints, etc. on the interior cell windows would permanently degrade laser performance as the beam had to "see" clearly through the windows.

lab with his air spray can in full operation. Burns got the hint. In spite of hating his stinky cigars, Cordy thought him a very pleasant man.

Kayak Sportsmen

Another of Korad's young engineers with an unassuming and sometimes deceptive appearance was Ben Parks, whose expertise in electronics,[79] along with that of Cordy, facilitated the production of the intricate circuits responsible for powering Korad's commercial lasers. (Cordy designed military electronics.) Parks had come from Florida, where he had built a reputation as a highly competitive canoe paddler. Having all but been raised in a canoe, he was able to perform maneuvers that left premier canoe paddlers staring in amazement. Cliff Cordy was Parks' partner on a number of outings and, having never seen the inside of a canoe himself, was amazed at Parks' abilities as an athlete. Although he was not a dominating physical presence, he was in good shape, having trained regularly at his sport.

Cordy had highly reasonable claim to fitness himself due to his penchant for running, which allowed him and Parks to share the sidewalks of Santa Monica on any given evening, enjoying their mutual interest in exercise as they loped along Sunset Boulevard and the adjoining foothills. One evening, as the two huffed their way alongside the front of a residence, they were startled by an abrupt ARROOF! ROOF! erupting from behind a three-foot-high perimeter block wall. Parks, who liked to boast of an affinity with animals, let loose with an ARROOF! ROOF! in return. The unexpected response, according to Cordy's panicked estimation, was the biggest St. Bernard ever born attempting—with near success—to launch itself over the wall. It was *Run Forrest, run!* as Cordy and Parks dashed to safety.

Tail-Gaters Beware!

The athleticism of canoeist Parks was subsequently demonstrated as he was spending a beautiful afternoon kayak surfing on the Santa Monica beach. He returned to his van just in time to see a young guy

---

[79] Ben Parks got an MSEE from the University of Florida and prior to Korad was working on a doctorate in plasma physics at UCLA, which he gave up for love of his work at Korad. He did all the electronic design, testing, and managed production of electronics for the commercial lasers. He was brilliant and my "electroniker" speaker at the popular Laser School I hosted for Korad.

making off with his camera. Parks sprinted after the thief, who was attempting his own mad dash, and ran him into the ground. As Parks stood over the younger man, who lay there gasping like a goldfish whose tank had been upset onto the living room carpet, he suddenly realized he was still carrying his kayak paddle. Thrusting it to within an inch of the offender's throat, he ordered him to calmly await the arrival of the police, who were not long in responding.

Arthur (Art) Lubin, 1977

Lubin, the Military Prime Mover

Art Lubin was a member of Korad's top management for three years until 1968 when the grand post-Maiman exodus took place. Lubin was as he appeared: exceedingly sharp in intellect and commanding in presence. He headed up the military group. After Korad, he founded the Laser Institute of America, LIA, based in Florida, and was its first president. At the LIA he recorded his former position at Korad as chief engineer. This he absolutely was not; Cliff Cordy, who worked for him, agrees and goes on to say that there were real engineers, such as Burns, Hoskins, and Pastor, all at his level of management. Cordy comments on Bill Buchman being the only engineer with a doctorate at Korad, one of few he knew in his life and the one who influenced him to get his doctorate after Korad.

Lubin was a first-rate manager. Cordy, who worked for him and thought highly of him, said, "his biggest contribution to Korad was that he knew what the military needed in the way of lasers, and could convince them that they ought to have those things. In short, he was a salesman. Prior to Lubin, Korad was an ongoing experiment. Lubin gave the experiment a sense of direction, at least in the military group." He landed government contracts worth a lot of money. During the Vietnam War, the military needed laser weapons desperately. They used Korad to quickly prototype working models, only to award production contracts to others. Which just goes to illustrate the old saying, *Pioneers get arrows in their backs*!

Hijinks

Another engineer known as quite a character among fellow Koraders was Lorand (Larry) Wargo. While his epic shouting matches in the halls with Art Lubin, who had hired him, served as entertainment for a time, Wargo was best known for his tall tales. One day he might claim to be a Hungarian freedom fighter, the next day the recipient of an Oscar.

Ray Larson, like Wargo, was resourceful. He drove the same model van as the GTE telephone service, conveniently given the same two-tone paint job as the telephone vans. Certainly it didn't say General Telephone on the sides, but everyone assumed it was a telephone truck because that was what they were used to seeing. Wargo would drive into the packed parking lot at Sears where an attendant promptly trotted over and waved him into a no-parking zone at the front door and set up cones around his truck. He thanked the clerk and assured her this would make the job much faster then he would hustle into the store without a tool in his hand.

Additional mischief was made possible by a switch Wargo installed under the dash of his van with which he could cut the power to his brake lights should anyone tailgate too closely. His van had a big, heavy, unforgiving, diamond-plate steel back bumper. He could slam on his brakes with impunity, and when it was over, take his foot off the brakes, turn the switch back on, and come gimping out of the van, practically crying in agony. "Whadda ya mean no brake lights? I have brake lights—see?

Dr. Fred the Party Animal

"Amongst Korad's top management, Fred Burns, second in command, was a life-of-the-party type. His daughter Nancy told me that in any party he joined he quickly became the center of attention. During one sales trip to Washington DC with Tony Johnson and Jim Kessey, Fred and I found ourselves at dinner in a Fourteenth Street Greek restaurant with a belly dancer as entertainment. Fred hopped up on the stage and took his shirt off. Appalled, in an attempt to distract attention from him, I quickly followed him on stage. The two of us with the voluptuous woman performed an exotic dance.

In the same year, 1967, we repeated our belly dancer act at a company sales meeting at the prestigious Paradise Point Resort in San Diego, this time with our wives in attendance. My wife said I was a funny man. We had been playing Bridge with Jim Kessey and his wife. Coming into the restaurant, I was cold sober, but seeing Fred Burns up on the stage making a drunken fool of himself, I ran up on the stage and began acting drunk while dancing with him. I was exceedingly pleased with myself because I had been a very shy boy. Fred was like a father to me. He mentored me on our sales trips together. I was shocked when he told me someday he would tell people he knew me back when. He also told me I had a unique ability to sell lasers that had never been built. His friendly encouragement meant a lot to me.

Need a Light?

Some pranks involved perfect strangers. After Korad's acquisition in 1968 of a low-power state-of-the-art red helium neon laser[80] that was to be used to align the optics of their large and more powerful ruby lasers, temptation beckoned. One could bounce the small and steady red beams off the two laser cavity mirrors to make them parallel so the laser would lase. Like mischievous children, three of Korad's marketing staff, headed by chief instigator Marv Sachse along with fellow hooligans Jim Kessey, Jerry Shane and yours truly, determined that a test of this new laser toy was essential. Night fell and we decided to see how far we could cast the beam and still see it. Hunkering down behind a room-sized window in Marv's office that looked out on Colorado Boulevard in Santa Monica, we began tracing lazy circles on the pavement and the opposite building. Directly across the street came a young woman. As she was attempting to unlock her car door, we illuminated the keyhole for her. Surprised, she looked down and around, but seeing nothing, quickly jumped in her car and drove off, leaving us howling like a pack of hyenas.

---

[80] Known as a HeNe (he-nee) gas laser, the original "laser pointer," was replaced by compact solid state lasers.

Yes, boys will indeed be boys, and what is the most enduring distraction for the typical boy but a pretty girl?

THE BALL: Korad Sales Crew Employs Innovative Methods

Back in the 1960s, during his days as a sales guy for Korad this author found it advantageous to carefully choose the proper place to entertain prospective clients. One particular establishment was only one mile from Korad's facilities at the northwest corner of Wilshire and Harvard in Santa Monica. It proved to be an ideal resource. Euphemistically termed a "gentleman's club," The Ball was an upscale nude entertainment club and restaurant that catered to male business executives. The sign out front read merely "The Ball," omitting any details as to the entertainers' state of dress or undress. It was a second-story walkup with outside stairs on the north side of high-rent Wilshire Boulevard. The owners of the high-end homes in the vicinity often voiced their distaste for the operation of a "strip club" in their neighborhood. The Ball was open until late at night, and the rowdy behavior of the gentlemen in attendance disturbed the locals' slumber. These residents campaigned against The Ball, resulting in a series of restrictions, but prior to that point I had used the place with amazing success.

Apparently when a man sees an attractive lady divested of her clothing, his frontal lobes (that portion of the brain responsible for higher reasoning) short circuit and he devolves into a grunting, panting, knuckle-dragging brute—no more responsible for his decisions than a man under any other intoxicating influence. In fact, a recent analysis of this phenomenon provided by Dr. Buchman, one of Korad's top engineering geniuses, included the following interesting observation regarding the dynamics involved: "It is my opinion that in spite of several centuries of enlightenment and recent political correctness, evolution has ensured that most men have the brains of male chauvinistic pigs. Women who realize that, and are able to manage the situation, can do very well indeed; while those who fight for political correctness are doomed to failure."

Elaborating on the neurological deficit men labor under, Buchman went on to say: "Men's brains are no more likely to change than women's. In that spirit I offer the analogy that pretty women are comparable to pretty laser resonators. Women with serious defects in appearance will not lose much in appearance from minor additional flaws. Similarly, a laser resonator with a serious distortion, say from a poor laser rod, will not suffer much from a minor additional distortion. On the other hand, really good-looking women—and I will use the young Elizabeth Taylor as an example—may have a minor flaw and it's really noticeable. For Elizabeth Taylor everyone noticed the mole on her chin, a flaw that would pass by unnoticed for most women. For an otherwise perfect laser resonator almost any defect will affect performance." My analytical colleague wrapped up his analysis by noting: "I have more experience with laser resonators than with women."

As an introvert and wannabe extrovert, I did not have the natural talent for sales that Korad's other marketing guys possessed, some of whom seemed born to it. Marv Sachse, sales manager at Korad, comes most prominently to mind. In his capacity as a marketing guy he spent a lot of time at The Ball, even dating one of the dancers for a time.

After insisting on seeing a photo of her and being presented with a nude one—the only one Marv had—Marv's father was horrified at his son's selection of a girlfriend.

I used to take customers there, occasionally for dinner but usually for lunch. It was a magnet and a major incentive to get them on a plane to visit us. As the product sales manager in the marketing department for industrial lasers, when I called a customer, I would advise them, "We can either eat in a nice restaurant or at this upscale club that features nude waitresses performing. If it offends you, obviously it's out, but I feel duty-bound to mention this place. I've been around the country a lot in my travels, and I've never heard of or seen anything like it." This line almost always aroused their interest.

I think there might have been one out of twenty-five customers who declined the invitation. For most men, it had the same effect as one would get waving a rib eye steak in the face of a starving hound. The Ball featured first-rate gorgeous girls, no sleazy fare. The nude performers were actually the waitresses who took turns dancing on small stages no more than five by five feet. Female patrons were even welcome, and once in a while a woman attended. The so-called sexual revolution, starting in 1960 with the advent of the Pill, was in full swing. Los Angeles was famous, or infamous, for its nude performers. Known as Sin City, or La La Land, it was a magnet attracting our customers to come for a laser demo.

The waitresses were merely topless because there was a city regulation stating they could not serve food if they were wearing absolutely no clothing. According to our Ball expert, Marv Sachse, it had to do with some sanitation thing.

Another restriction enacted by the city of Santa Monica was the requirement that the performers had to be eight feet away from the spectators. So the owner, a rotund fellow who went by the name Whitey, erected four-foot-high plastic walls around the performers' elevated stages. This sufficed to satisfy the new requirement because, as the bug crawls, that involved four feet up the wall, and four feet down the other side.

I tried to be a thoughtful host by always taking the seat facing away from the stage, which seemed to irritate the girls as they felt I was ignoring them, which I was. I was concentrating on the customers and their reactions, and I enjoyed that while keeping my mind on my business, which was selling by entertainment and relationship building. The girls came to take great delight in tormenting me, often stalking off their little pedestal stage completely naked to approach my table where they proceeded to sprinkle talcum powder over my head. Another favorite tactic they employed to get my attention was to walk up and put their panties over my head. I made a show of standing up and entertaining the whole restaurant all the while loving the opportunity to show off. A third measure employed by another young lady involved removing my glasses and placing them over both her breasts, after which she took a deep breath and broke them. That particular ploy bothered me, but I swallowed my irritation and wrote off the broken glasses as a business expense. Naturally, I told my boss, Tony Johnson, about it and secured approval for putting it on my expense report. He loved the story and so did all the other male Koraders. Perhaps it amused the female secretaries but I never asked. My stock as super salesman rose, which pleased me greatly. This place was super good for sales at that time and place.

Accounting Sees Red, Then Green

My dedication to sales landed me in trouble with Korad's accounting department, which felt my regular visits to The Ball were financially prohibitive, if not ethically and morally. But accounting could not argue with the resulting success as orders flooded the factory. Accounting has little power in small high-tech manufacturing companies;

engineering and sales are the top dogs. Sales boss Tony's general restaurant rule was for us to avoid any discussion of business until after the meal, when the client could concentrate on the sales message. The Ball modified Tony's rule somewhat; we did more relationship building. For example, I told stories about talcum powder, panties on my head, and breaking my glasses. Tony understood very well a man's mind cannot be on business when dancing naked females are so close. The Ball was a wonderful tool for lowering sales resistance.

The Ball was a very popular place and always close to a full house. With the obvious exception of the waitress/dancers, back then the male customers all wore suits and ties. We used to speculate about what an enviable job Whitey had interviewing girls for his club. He was a laidback guy who loved to circulate and schmooze with his clientele; he seemed quite at home in his role as host. He had another place out in San Fernando Valley, appropriately named Whitey's Other Ball.

Korad's industrial laser customers especially loved the place. A lot of orders were received—over one hundred as I recall—at $25,000 to $30,000 a pop. That was real money in those days—$3 million—when the salary of a small company president was around $25,000 a year and the cost of the standard three-bedroom, two-bath home in the area was about $25,000 as well.

There were, however, dissenters, like one gentleman who accompanied Korad's Stan Parnas. This client had not been advised as to the specific nature of the entertainment available, and it turned out he was Mormon! Extremely offended hardly sums it up, and that particular deal instantly evaporated.

Recollections of The Ball from Koraders interviewed for this book resulted in some additional observations. Dr. Jim Boyden, Korad's engineering manager, recalled a client from Idaho who was so impressed with his experience at the club that he returned the favor by taking Boyden prairie chicken hunting on one of the latter's business trips in that direction. (This is not a male chauvinist euphemism for a female human; these are actual fowl.) Boyden scored one bird, which was partially frozen that night by his host, only to thaw thoroughly before the flight home concluded, which made Boyden unpopular with his fellow passengers as well as his wife upon his return home.

And then there was the memory, shared by several, of Korader Jim Kessey struggling to maintain his composure while tying a bit of yarn around an accommodating dancer's nipple. Super salesman Jim has been one of my very best friends ever since Korad and was my sales manager at a laser company I co-owned for twenty-two years, Florod Lasers. Florod was a Korad competitor. In fact my first purchase order was for an laser at RCA New Jersey that I took away from Korad.

The Ball was used as a company sales tool for ten years between 1965 and 1975, during the tenure of four of Korad's chief executive officers: Maiman, Zerby, Thurber, and Don Sims. Buchman summed up his feelings about those experiences, musing in a recent e-mail to the author: "I suggest that next time we have a Korad gathering, we invite The Ball performers to it. It would be interesting to see how they stood up to the test of time."

Chapter Nine: Maiman Leaves Korad

Mike Weiner, graduate of University of Southern California in Los Angeles, the crosstown rival of UCLA, had bachelor's degrees in mechanical engineering and business, and a master's in business administration (MBA). He came to work at Korad in 1967 and worked there until 1977. He recalls an atmosphere of "We can do anything and we can build anything." He reflected in our interview about Korad's lasting legacy as a womb for entrepreneurial creativity and innovation, while acknowledging its weakness in day-to-day operations. "Indeed if you were trying to have a company that was going to make a lot of money and leave a long legacy as a fine manufacturing company, Korad was not the model for it. But Korad might have been the right model for an entrepreneurial company that would try everything and be a very good incubator of product ideas."

My experience at Korad was frequently frustrating. I recall being particularly irritated in an attempt to get cost information from the head of manufacturing, Hal Moss. "What do you mean you don't know the cost of the K-RT Resistor Trimmer[81]?" I once asked him. "I've sold—and you've shipped—fifty of them."

[81] Korad's most successful industrial product, the K-RT, sold more than all competitors combined in the early 1970s. First order came from Don Yoder at GE Schenectady, New York, 1968. General Motors 1970 model year car radios were trimmed by CO2 lasers. I succeeded in replacing them with YAG, one of the major successes of my career.

He replied, "Rod, go see accounting." With accounting, I got nowhere. They were a stone wall.

From Ed Young, the big boss of accounting—and the company's money—and controller at Korad from October 1965 to May 1968, I subsequently learned that Hal very well knew the cost. I suspect he was marking up the cost so as to increase profit; and then, no doubt, he just wanted to get me out of his office.

Gossip Can Be Wrong

The scuttlebutt in those days was that Korad was Carbide's little hobby. They were dumping money into it but were unconcerned as to whether it was profitable; Carbide didn't want a profit. They just wanted a place to stick money so that they didn't have to pay the IRS. It seemed at times as though it was a sinkhole for excess profits in the corporation. They could avoid paying taxes on it while using that money productively to investigate the viability of money-making laser applications. For us, that was good indeed.

But Ed Young said the gossip mill was dead wrong; Carbide did care a lot about profitability. In fact, in 1967 he enabled Korad to make the highest profit in the entire electronics division of eight departments. This giant feather in his cap made him the golden-haired boy and a hero to management, thus establishing his successful Carbide career and retirement.

Korad Worth Up to $4 million!

Ted Maiman was very interested in acquiring Carbide's 80 percent ownership of Korad, according to Young, who speculates that such a move would have cost Maiman in the neighborhood of $3 million to $4 million in 1967.[82] He and Maiman had worked together smoothly. "I can't comment on what kind of a scientist he was, but I can tell you he was an excellent businessman. He was aces as far as I was concerned."

Recalling his arrival at Korad, Young said, "I knew that any subsidiary that wasn't wholly owned, that there was going to be a certain amount of potential clash with Union Carbide." So he decided

[82]$4 million in 1967 was worth $27.9 million by 2013. (US Inflation Calculator on the Internet) Korad's 1967 revenues were over $5 million a year and the company employed over 100.

that the only way this was going to work out was if he forgot all about having worked many years for Carbide. He told Maiman, "It is my intent to work 100 percent for Korad and do what's right for the business. I'm not here as their representative. I'm here to make this business succeed, and if it does, not only will your team be winners, but Carbide will be an even bigger winner." Maiman was impressed and the two became instant friends.

Young praised Maiman as one of the best managers he had encountered in his thirty-plus years at Carbide. Maiman in his role as businessman dived deeply into the company's financial records. Young said, "One of Ted's pet peeves, which he made me aware of almost immediately, was the number of mistakes that got made during the month by the scientists and technicians who were working on projects. They charged parts from the stockroom to the wrong account so labor on the timesheet would be to the wrong account."

As a result, Maiman was greatly annoyed at not knowing how much it cost to make things and therefore could not accurately set prices. So he asked, "Ed, can you devise a way to make all of these corrections *before* the books are closed instead of *after* they're closed at month end, so they're all in the right month?"

It took Young about two months to set up a more effective and accountable system, referred to as The Flash Report. Maiman was pleased with his new accounting system, which enabled him to more effectively run the business.

The Other Shoe Falls

Toward the end of the Maiman Era in 1967, when Carbide was preparing to exercise their final takeover option, Young recalls Maiman's concern: "Ted asked me, 'Do you think they're actually going to exercise that option? I don't know. I would really love to own this place.' He wanted to know what I thought the chances were that Carbide would pass the option, or else give him the option." Young was unable to encourage his friend, saying that in his opinion such an outcome was unlikely, to which Maiman responded: "Well, I don't want to work here if they're going to exercise that option."

Unfortunately for him, Carbide did acquire Korad in its entirety. In the fall of 1967, Ted Maiman, the inventor of the laser, departed the company. The decision sent shockwaves throughout Korad, as one after another of its key players announced their resignation soon after, with virtually the whole team and all the top management leaving within half a year.

Ed Young didn't want to be at Korad after Maiman so he requested a reassignment from Carbide and got it. Sales manager Marv Sachse recalls quitting his Korad position in November 1967 due to Carbide's acquisition. Sachse had just won for Korad the prestigious IR 100 Award in New York City and remembered Maiman's instructions to him prior to the latter's departure from the company. "Maiman called me into his office and said, 'I want you to get the IR 100 Award, and I want it stated that the award went to Korad Corporation of Union Carbide, not Union Carbide's Korad!'"

Maiman's anger at Carbide existed alongside an agonizing antagonism toward Charles Townes and Arthur Schawlow of Bell Labs regarding credit for the laser's invention, which we touched on in an earlier chapter.

Controller Young recalls Maiman's frustration: "He was involved in litigation at the time with Schawlow and Townes, and I never knew the particulars." Between Maiman and Townes, in particular, there was a lot of animosity. Young called it "a lot of grinding of teeth," and went on to say, "It came up in casual ways. Sometimes in his office, I would say, 'Ted, you look a little troubled today.' And he'd reply, 'Those damn guys.' Then he'd tell me about what the latest was, but it did trouble him a lot. He wanted recognition for his creation of the laser."

Venture Number Two, Maiman Associates

Korad was Maiman's first venture into the business world as an owner and a company president. His second was a consulting company, Maiman Associates, founded in 1968 and specializing in management and start-ups. Maiman was the sole owner. According to Ed Young, his controller for three years at Korad, he hoped his reputation would attract business, but it didn't work out. He started out big when he should have stayed small, Young told me in our interview on March 26-27, 2010. Maiman's rented office space was in the thousands of square feet—too much space—and located in the high-rent district near Wilshire Boulevard in the Beverly Hills area.

Ed Young resisted Maiman's attempt to lure him away from Union Carbide. "He wanted me to be his financial director and work with his clients in order to develop business and financial plans." Maiman, Young noted, was persistent to say the least, making repeated telephone pleas. Young recalls one particular exchange. "Ted, we've been over this a dozen times. I'll give you any kind of help you want, but I've been with Union Carbide for seventeen years. I can't really throw that away." His three years at Korad was included in the seventeen.

"Well, I want you to think about it," Maiman said. And Young did contemplate the move, only to decline even after an arm-twisting move by Maiman, who brought him to his skyscraper office near prestigious Wilshire Boulevard, where there was a spectacular view of downtown Los Angeles.

"When I saw that office, it was right after he leased it. Nothing was happening there yet; he just wanted to show me. It was part of his sales pitch to get me to stay."

Venture Number Three, IDAC

Marv Sachse recalls a phone call from Maiman after Maiman Associates had failed. "He contacted me regarding IDAC, a telephone communications company that he had just bought. But he really only had concepts of products; there were no finished products. They weren't doing manufacturing. They only had ideas." Sachse discussed the difficulties they faced in having to literally redesign everything. IDAC's primary offering was a telephone dialer, and there was a good potential market for such a telecommunications product, such as in airports to call hotels for reservations. "I think Maiman surrounded himself with people who compensated for his deficiencies...you know, made a good, synergistic team."

Maiman attempted to draw from his primary colleagues at Korad when undertaking his new ventures. Sachse recalls buying a VW van and departing for a lengthy skiing trip across Europe after leaving Korad. His plans were interrupted by Maiman's call. The buttoned-up and reserved laser inventor was taken aback when Sachse returned for a meeting. "When I came back from Europe, I showed up in a tie-dyed tank top and hippie beads and a beard. Maiman just about freaked!" In spite of that, Maiman hired Sachse, knowing very well his sales talent.

Sachse, Maiman's sales manager at Korad and at IDAK, told me in our interview of June 30, 2010, that the company was founded by Maiman with venture capital backing. Maiman was administrative and not involved in engineering. IDAK failed, according to Sachse, because too much money was paid for a supposedly viable company which turned out to have a terribly incomplete design; it was not ready for sale. He said the company ran out of funding, the venture capitalist gave up, and that this venture lasted only nine months in 1970.

Venture Number Four, Laser Video

After his IDAC venture, Maiman took another stab at entrepreneurship with a start-up business involving laser-based color television sets. I have not found anything written on this, his last venture, so I feel duty-bound to tell the story. His concept was technically simple: take three lasers—red, green, and blue—called RGB, and combine the *continuous*, not *pulsed*, beams into a color television display.

Laser Video Corporation was cofounded in 1971[83] with Hal Moss, his former manufacturing manager at Korad. He invited me over to check out his new laser baby, so I visited him. I was impressed with the brilliant colors but not with the blurry images. He allowed as how this was a problem he was currently working on. But the fuzziness turned out to be a killer. It was caused by laser "speckle," a side effect of the super-high coherence characteristic of lasers. In science and medicine, knowing the cause sometimes leads to a cure. Today, laser TV is a reality in high-end TVs and movie theaters. The lasers used now are low-power, solid-diode types of lasers that are related to the semiconductor lasers used in computer memories, DVDs, and laser pointers.

Business Failures are Excusable

While some were quick to lay the blame at Maiman's feet for his three business failures, others who had worked alongside him for years had a more thoughtful view of the outcome. Controller Young said, "There's probably only one out of ten or twenty new ventures that are started up and actually succeed, so it doesn't surprise me that Ted was involved in a few flops. If you don't have the right idea, the right people, or the right product, or can't get it across to your customers in some way, you tend to fail. I wouldn't necessarily conclude that Ted was not a good businessman because after he left Korad he had some failures. Sometimes it's the idea that's flawed."

---

[83] "Maiman, Founder Laser Video," *Laser Focus*, (Aug. 1971): 48.

Life after Failure Goes On

In 1976-77 Maiman joined an established firm, TRW Electronics in Redondo Beach, California, where he was vice president for advanced technology. He insisted on reporting only to the big boss, Dr. Simon Ramo, the R in TRW.[84] Maiman was with TRW for twenty-three years until his retirement in 1999. His advanced managerial career spanned thirty-eight years, from 1961 to 1999.

His first marriage to Shirley Rich began in 1956 and ended thirteen years later. They had one child, a daughter, Sheri, who lived only thirty years, died of cancer, married but childless. This was the tragedy of his life. Quoting his memoir, *Odyssey*, "I can't say that I have ever been able to truly recover from that tragic blow."[85]

The light of his life was his wife Kathleen. His description of her in *Odyssey* is priceless: "As I got to know Kathleen, her creativity, her active enquiring mind, her sense of humor and her admirable sense of values, she became more and more irresistible."[86] They had a happy marriage. I admire her making it her life's work to continue his fight for recognition as the "creator of the world's first laser" as he calls himself on the cover of his *Odyssey* book.

They met on an airplane when he was returning from having been inducted into the prestigious National Inventors Hall of Fame in Washington, DC. Quoting from *Odyssey*, "Of the more than five million United States Patents issued at that time, only about fifty inventors had been inducted [into the Hall of Fame]."[87]

Maiman was an inventor of more than the ruby laser. According to Dr. Bob Charpie, in addition to patents on the first laser, he held patents on masers, laser displays, optical scanning, and laser modulation.

Maiman's Bio Summarized

Education:

- BS Engineering Physics, University of Colorado (1949).
- Columbia University (transferred to Stanford).

---

[84] Tidbit on Maiman insisting on working only for TRW's big boss is thanks to Kathleen Maiman in a personal communication.

[85] Maiman, *Odyssey*, 194.

[86] Maiman, *Odyssey*, 176.

[87] Maiman, *Odyssey*, 174.

• MS Electrical Engineering, Stanford University (1951).
• PhD Physics, Stanford University (1955).

Experience:
• TRW, VP Advanced Technology, 1976-99.
• IDAK, CEO, 1970-1971.
• Maiman Associates, sole owner and CEO, 1968-69.
• Korad Corp., Founder and CEO, a partially owned subsidiary of Union Carbide Corp., 1962-67.
• Quantatron VP, 1961-62.
• Hughes Aircraft, Physicist, Hughes Research Laboratories, 1955-61 (According to *Odyssey,* ended as Section Head of Quantum Physics.)
•

Awards and Honors:
• Japan Prize, 1987, the Japanese "Nobel Prize."
• Israel's Wolf Foundation, Wolf Prize in Physics, 1984.
• Fannie and John Hertz Science Award, 1966.
• Member of the Board of Control Laser Corporation.
• Franklin Institute Ballantine Medal, 1962.
• American Physical Society (APS) Buckley Condensed Matter Prize, 1966.
• American Optical Society Wood Prize, 1976.
• Authored autobiography, *The Laser Odyssey,* 2000 (according to his wife, Kathleen, unlike his rival, Charles Townes, he had no ghost writer). She watched him write it and I viewed his original handwritten manuscript at her residence in Vancouver, Canada, on July 29, 2013. An excerpt is available at www.maimanbook.com.

Ted and Abe Maiman, father and son, about 1985

Maiman's Early Life

Theodore Harold Maiman was born July 11, 1927 in Los Angeles to Abe and Rose Anne Maiman. In *Odyssey,* Maiman says Abe was "a warm loving father" who kept a small electronics lab in every home they had and taught his son electronics. At age twelve Ted got an after-school job repairing electrical appliances in Denver. At age thirteen, at the beginning of World War II, he was running the repair shop entirely on his own after the owner took a job in a defense factory. However, the owner would pay Ted only twenty-five cents an hour, so at age fourteen, Ted got another job with a forty percent raise also doing electrical appliance repair. He kept this job in high school. After dinners at home he took a night class in radio theory at the University of Colorado.

In 1943 at age fifteen, he moved from Denver to Morristown, New Jersey, where his father had been transferred from AT&T's Mountain States Telephone & Telegraph Company to the famous and prestigious Bell Telephone Laboratory in Murray Hill, NJ. Ironically, Bell Labs was to become his chief enemy in his struggle for recognition as the inventor of the laser.

After graduating high school at age seventeen, Ted got a job as an engineer designing military test equipment and power supplies. His boss, Ed Hamilton, took him under his wing and taught him even more good electronics stuff. The two of them coauthored a technical publication, Ted's first.

Maiman the Sailor age 18[88]
1945

Besides getting his first job as an engineer in his seventeenth year, he passed a national test for a first-class commercial radio license, the sole requirement to work as a radio station engineer. He was the youngest person in the United States to hold this license. Before his seventeenth year was over in 1945, he enlisted in the U.S. Navy and became a radar tech.[89] Radar experience was instrumental in his invention of the first laser. (Indeed, knowledge of radar was a requirement for a job as a technician at Korad—reference the Hal Walker story.)

I suspect the postwar GI Bill helped finance Maiman's education, as his mother's illness had sapped his father's resources. She was a somewhat unbalanced personality who wanted her son to be a medical doctor. She never accepted his doctorate in physics.[90]

---

[88] Reproduced with permission of Kathleen Maiman, and taken from his personal papers. This applies to Father and Son, Sailor and Machinist photos.

[89] He served in the U.S. Navy for one year, 1945-1946. A copy of his discharge is available at www.maimanbook.com

[90] Personal communication from Kathleen Maiman.

Maiman the Machinist-1995

A Many Talented Man

Totally unknown to the world is that Maiman was an accomplished machinist. Shown in the photo above is Maiman in his garage machine shop. The reader may recall the story of his Stanford classmate, Ralph Wuerker, the master of many trades. Now we know that Maiman was cast in the same mold.

Like a diamond, Maiman had many facets. A listing of the many faces of Maiman is in the Index under "Maiman". As Ed Young, his controller and lifetime Carbide careerist said, he was a master businessman. He did what experts in business do: he studied his financial statements and ran the company accordingly, namely to decrease costs and increase profit.

Marty's Stories

In an interview with Korad's personable and quite technical saleman, Marty Phillips, on April 12, 2010, he related how, as a man in his early twenties in 1964, he was hired by Korad as an applications engineer. Armed with a bachelor's degree in physics, he was well equipped for the job. He was highly articulate and personable, charming even—ideal qualities for a salesman—so he was quickly groomed to be the company's East Coast sales manager. The relocation suited him and his wife Gayle just fine as both were from New York. He was a graduate of the Bronx Science High School, the top of the heap. Such a fast promotion to management was and is highly unusual, but then, no one had laser experience at that time and us young guys were fast learners and besides, we were paid less than the older chaps.

For training purposes Phillips was invited to join Maiman's entertaining of visiting dignitaries. He recalled how lots of top professors and other scientists wanted to meet Maiman, the famous laser inventor. He was like a magnet drawing influential people who controlled funding of their projects. He went on to say that his sales abilities motivated these scientists to buy Korad lasers, the most expensive on the market. Phillips added, "He had his fingers on the key technical issues and products that we were coming out with. It was very informative. I learned a lot from Maiman."

Probably the most memorable were two stories Maiman told at many of their customer luncheons. Charles H. Townes, inventor of the maser, came to visit Hughes Research Labs in Malibu where Maiman was diligently trying to make his ruby laser work. Townes succeeded in convincing Maiman's management that ruby would never lase, so they tried to cancel the project. Working night and day he and Ray Hoskins got lasing.

The second story was that Bell Telephone Labs (BTL) had many of their people sitting on the Physical Review Letters (PRL) review board. Maiman blamed the BTL scientists for killing his announcement of his invention of the first laser and refusing to publish his paper. That's why Maiman had to publish in the British journal *Nature*. These two stories, told many times by Maiman to visiting scientists, underscore how Townes and BTL were his lifelong enemies.

On May 5, 2007, in Vancouver, Canada, where he lived with his wife Kathleen, Ted Maiman died at age seventy-nine from systemic mastocytosis. His ashes are with his daughter, Sheri, in Los Angeles.[91]

---

[91] Verbal communication with Kathleen Maiman.

Chapter Ten: Diamonds - Improving on Perfection

This story from the early days of the laser is a timeless tale about diamonds, which are said to be a girl's best friend. It's a story for every woman who loves diamonds and for every man who loves a woman and buys her a diamond. Floyd Pothoven and I claim to be the pioneers in laser drilling of diamond gemstones and the first to discover that you can actually blow up a big stone (two carats!) with a laser.

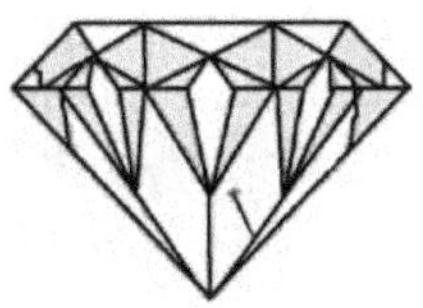

The year 1968 found Korad steaming forward with those who had remained after the Maiman exodus and with some fresh faces, including Maiman's replacement, Dr. Clayton Zerby. He was a brilliant mathematical physicist while a bit of a cold fish in his interpersonal relations. I Googled him and found online a copy of a study he did on

[92] Clayton D. Zerby and Henry Brysk, "Theoretical Calculations of the Bremsstrahlung Cross Section," June 24, 1966, Work Performed Under Contract No. NASw-1235, NASA.

atomic interactions.[92] I was impressed by how heavy in math it was. He and the boss of the electronics division, Bob Charpie, had worked together at Carbide's famous Oak Ridge Nuclear Lab, the place in Tennessee where all the weapons-grade uranium for the Hiroshima A-bomb had been produced and basically the place where World War II was won!. I was surprised to learn from Dr. Charpie that he had been Zerby's doctoral graduate advisor at the University of Tennessee, where he was an adjunct professor of physics.

Clayton (Clay) Zerby, 1971

Laser Diamond Drilling

As a good and loyal Carbide company man, Zerby was anxious to make his mark as the new head of Maiman's Korad. An opportunity came to help diamond dealers improve the quality of their cut stones. The crystal products department of Carbide's electronics division was supplying artificial sapphires and rubies to the jewelry trade, which were used for cutting facets in gemstones. Profits were exceedingly high for all concerned and Charpie and Zerby were eager to help their friends make even more money. They had established relationships with the jewelry trade and planned to expand on this asset. After all, sales, as well as life, are all about human relationships.

The high-tech nature of Korad's work never made for a dull day; still, Floyd Pothoven, who was to become my partner in a company, Florod Lasers, which we co-owned for twenty-two years, was an engineer at this time. He and I were surprised and intrigued by the unique request of Korad's latest client, renowned Manhattan diamond dealer Lou Perlman, who, along with his brother, Max, presented us with a rather curious challenge. It seemed that many of the stones they initially obtained from the DeBeers diamond mines in South Africa contained one or more minute flaws. Not cracks in the stone itself, but mostly just small particles of carbon that had not crystallized when the gem was formed deep in the Earth millions of years ago. These inclusions, or ***piqué*** as they are known in the trade, interrupt the various rays of light reflected from facet to facet, reducing the brilliance and thus the stone's value.

Could Korad, Perlman mused, conceive of a way to somehow access these flaws, allowing for their removal? He had heard about Western Electric Division of AT&T having drilled industrial diamonds so that wires could be pulled through the diamond "donuts" to make ultra thin wires. Perhaps the Korad's minutely focused lasers could be used to bore a small and hopefully invisible hole until the flaw was reached? Imagine the potential to vastly increase the value of these gems should such a procedure prove to be viable! Drill a hole so tiny that nobody could see it—what a juicy opportunity!

These fine holes would get down into where the tiny carbon deposits, the *piqué*, were located. The idea was to create tunnels barely visible with a jeweler's standard 10X eye loupe then stop at the carbon deposit, an imperfection that reduced the diamond's sparkle (called "clarity" in the trade). Next, if necessary, we would go into the tunnel with acid to bleach out the carbon deposit so as to make the diamond more perfect, have more sparkle, and be worth more money.

Understandably, Perlman didn't want to ship his precious gemstones to Korad through the mail; instead he opted to carry them on his person as he flew from New York to Los Angeles. When he finally arrived at Korad's Santa Monica plant, I was there to meet him because I was the sales guy tasked by Korad with looking into this challenge. Although I was only thirty years old at the time, the fiftyish Perlman and I hit it off immediately. The gem man's short stocky build and round, balding countenance reflected a pleasant and jocular manner. Diamond dealers necessarily traveled light because they feared being robbed. Furthermore, the valuable stones were sewed into Perlman's long black coat. Hidden under his clothes, he sported a carefully cinched money belt. He told me he and his colleagues had no luggage, not even a carry-on.

Check This Baby Out!

Slipping a finger into a vest pocket, Perlman produced a hulking, two-carat stone in a beautiful gold setting. This beauty was easily a quarter of an inch across but unfortunately bore a rather large inclusion. Pothoven carefully placed the gleaming stone under the company's laser driller whose active element was a synthetic ruby rod. Could the laser's highly focused one joule[93] pulse be adjusted in intensity so that it vaporized a minute portion of the diamond with each blast without fracturing the gem? Would it be possible to thus bore a tiny, hair-sized tunnel from the bottom of the stone and reach the carbon particle lurking within? Diamond drilling with lasers for blemish removal had never been attempted before, and the tension in the room was palpable as the laser was fired.

Nothing! Not even a scratch was evident on the gem's surface. Pothoven increased the laser's power to near its most intense output before cutting loose with another blast. But still . . . disappointing. We were almost prepared to concede that laser technology might not be an effective means to remove inclusions from these stones when an idea struck me: perhaps just a dab of printer's ink on the diamond's surface at the point where the laser was to begin its cut. After all, printer's ink is carbon-based. Could it absorb just enough laser energy to initiate the drilling process? Pothoven carefully used a Q-tip to apply a dab of ink to the gem's surface. The machine was again fired, and...Yes! The diamond now bore a tiny abrasion where the laser had inflicted its first dimple. Firing at one blast per second, the process proceeded slowly, demanding patience. After a few minutes, we concluded that the process was too time consuming at its present rate, so we dialed the laser power to maximum.

Ooops!

The stone shattered into hundreds of tiny fragments. A shocked silence gripped the lab as Pothoven and I glanced anxiously at one another, steeling ourselves for the inevitable screams of despair. But Perlman simply shrugged it off. "Forget it, not to worry. That was nothing!" he scoffed. "Let's get on with it!"

---

[93] Even though a joule is only a quarter of a calorie, when you focus that to a microscopic spot, you get a very high density of energy.

Having recorded our measurements as the laser power was increased, we were able to determine that the stone shattered under the application of one joule of energy. Better to stay under one joule and hit the gem with hundreds of individual blasts rather than rush the process and end up with this day's result, we reasoned. The price of this initial lesson became painfully evident. Subsequent attempts on another stone met with a more satisfactory result, finally producing a tiny hair-sized tunnel directly to the inclusion.

We then watched with fascination as Perlman forced an acid solution down the hole, resulting in the elimination of the carbon offender. After rinsing out his bleach solution, he then replaced it with a refraction-matching liquid. This was quite effective in hiding the laser drilled tunnel from view by the average buyer, who understandably lacked the expertise necessary to examine the stone for evidence of such tampering. Diamond sellers to this day engage in this ingenious technique, which can mislead vulnerable buyers.

Solid lasers like ruby and later, YAG, were naturals for treating diamonds. Why? The temperature at the focal point of the beam focusing lens was thousands of degrees Fahrenheit, high enough to vaporize diamond but in such a short time interval of billionths of a second, that there was no time for the heat to diffuse into the stone which remained at room temperature during drilling. The high temperature and no heating were a charming combination.

Pothoven's and my pioneering work at Korad allowed for this first use of laser drilling to enhance the appearance and raise the value of gemstone diamonds. Carbide was pleased with our diamond project.

Diamonds, Inc.

I subsequently traveled to the diamond man's Manhattan headquarters on West 47th Street between 5th and 6th Avenues in the famous Diamond District. Buzzed in the front door, I found myself in a small antechamber about eight feet square and devoid of furniture or windows. With only a solitary surveillance camera to keep me company, I received a sudden start when a voice out of nowhere asked me for my name and the nature of my business. After dutifully answering to the satisfaction of the disembodied inquisitor, I apprehensively entered an altogether ordinary-looking lobby, complete with female receptionist. After a brief wait, Perlman appeared and engaged me in small talk, noting that his operation was staffed exclusively by direct relatives. "Like distant cousins even?" I probed.

"We don't go any further on the family tree than first cousins."

Glancing around the facility at the myriad male minions scurrying about, it occurred to me that Perlman must have a sizable family indeed. The trip was a success in that Perlman agreed to the purchase of a laser to the tune of $30,000, which was paid in cash—300 one-hundred-dollar bills. Thirty grand in those days was more than a typical small company president made in a year.[94] This made quite an impression on me, as I brought home only about a third of that figure annually.

Floyd Pothoven, 1974

A second trip by me to Manhattan was required for setup and training purposes after the laser had arrived at Perlman's Manhattan jewelry shop. A second laser, subsequently purchased by Perlman for installation in his Belgium plant, was installed and set up by my colleague, Floyd Pothoven, who was given a two-carat, undrilled diamond for his trouble. Floyd was the quintessential midwestern engineer—all business, and looking the part with his stocky build and distaste for small talk. Still, I believe I detected uncharacteristic buoyancy in his manner when he returned with that diamond. Just imagine how much I wished that I had been selected for that particular assignment!

---

[94] $30,000 was a typical price for a nice three-bedroom, two-bathroom house in the West Los Angeles beach communities.

After his first trip to install the laser, the diamond dealers in Antwerp, Belgium, had experienced some difficulties with their laser driller. Pothoven was tasked with flying a second time to their location to make it work again. "It was quite a trip," he reported to me, "I remember seeing these Jewish guys with their big buxom Swedish blondes who were a lot taller than they were." Pothoven recalls super-palatial city apartments and wild restaurants. "One of them was like an old-time wine cellar. You'd wind down a scrawny, little alley full of garbage cans and trash, then down the stairs, and then all of a sudden you're in this huge wine cellar that was just outfitted to the T's. It was incredibly beautifully furnished. The prices were just totally outrageous!"

Pothoven soon discovered the cause of the laser's trouble: a fried 2N3904 transistor. The diamond merchants were soon up and running and making an obscene fortune by removing inclusions from their stones. "Every so often, the laser followed the natural crystal structure or bounced off an invisible crystalline defect in the diamond, and it would actually curve around, drilling a J-shaped hole. It would actually come right back up toward the top again, just curve around and make a U-turn," Pothoven explained. Further, it was not always the best course of action to drill the stone from the back. "A lot of them we drilled in the face. It depends on where the defect was. You always wanted to go the shortest route. It didn't matter at all whether you drilled from the top or bottom. If you could keep the hole down to a thousandths of an inch it was not a visible hole. And if you could get rid of the black imperfection, that small hole was insignificant compared to what you had done to increase the value of the stone."

Scrubbing out Carbon Bubbles

Drilling to the defect was only half the battle, however. It still frequently needed to be rinsed out, and Pothoven recalls seeing that process in action. "In Antwerp, they had hot plates with acid in a fume hood, just boiling away. I'm not sure what the liquid was that was boiling. They didn't want to talk about their trade secrets." If they were lucky, the laser would vaporize the black carbon imperfection.

Korad's contribution to the practice of diamond drilling for *piqué* removal ended at this point as other companies with faster YAG lasers, like the ones employed by the Raytheon Corporation of Boston, Massachusetts, drew the gem industry's attention. Raytheon could deliver a couple hundred laser pulses a second compared to Korad's puny one pulse. Even this impressive performance was shortly eclipsed by Lee Laser of Orlando, Florida, which boasted an incredible 3,000 laser pulses per second in addition to an enormous increase in power.[95] This fabulous capability, coupled with relatively low costs, high reliability, and ease of maintenance, won the hearts (and pocketbooks) of those in the diamond industry.

As I have come to realize, too few jewelers disclose to their clients when such an alteration has been made to a stone; as a result, I strongly advise buyers to be cautious when shopping for gems. I once asked a jeweler if he carried any laser-drilled stones, which drew a hasty and heated denial. The next moment, the merchant began scurrying about his shop, gathering up the bulk of his loose stones, declaring he was closing for the day.

One need not take the jeweler's word that a particular stone has not been thus altered; one can simply request that the dealer loan him his 10X eye loupe, which will allow for a revealing inspection. Under this degree of magnification, wider tunnels are visible but narrow ones are not. Unfortunately, such an inspection is impossible for holes drilled from the back if the stone is already set, so it behooves a buyer to ask for a written guarantee that no laser drilling had been conducted, or better yet, get it graded and appraised. Buyer beware!

Not only does drilling make the diamond less valuable, but the process does not always go off without a hitch. Natural diamond is full of defects in the crystalline structure, and therefore the direction of the tunnel can veer off in any direction in a "J" curve, just as Pothoven described. What is our resourceful gem dealer likely to do to address this problem? Why, simply drill another hole! With two holes the stone's value is further diminished, but the hapless buyer is unlikely to notice, so again… buyer beware! Actually, a member of the staff at the Gemological Institute, Robert Weldon, told me he had seen stones with hundreds of laser holes in each one! Incredible! *

---

[95] Power in this context is Q-switched peak power, the power of a single laser pulse, about 20 kilowatts. The big deal about lasers is their ability to drop all that power into a small spot, in this case a third of a hair's width. Peak power densities of a billion watts per square centimeter are routine, enough to vaporize the hardest natural substance, diamond.

Caution is all the more necessary today because contemporary technology can supply UV frequency tripled lasers, where one can drill holes invisible to the jeweler's eye loupe but not invisible to a high-power microscope. However, you won't find such a microscope in a jeweler's shop, and you cannot easily carry one about. The solution is again to get it graded and then appraised, ideally by independent company(s) not affiliated with the jeweler.[96]

Besides J-curve double holes, multiple tunnels and invisible UV laser drilled holes, filled holes are common. These are virtually invisible by virtue of the filling being transparent and made of a index matching material so light bouncing around between the facets passes through the filling as if it weren't there – therefore totally invisible which increases clarity (sparkle) and value. So if typically glass filling were to fall out, there goes clarity and value. On the plus side a market exists for filled diamonds because the clarity is excellent. However, full disclosure is ethically required by the diamond merchants.

Disclosure and Fraud

According to an article in a jewelry magazine,[97] "Because of their potential to deceive, gem treatments, including those applied to diamond, have long had an aspect of fraud about them. Gem treatments, it must be recognized, are neither good nor bad in themselves—fraud comes about only when their presence is concealed." Some dealers disclose, and some don't.

So, the thing is this, laser diamond drilling has been with us for nearly a half century. The diamond industry has failed to enforce disclosure of such drilling. A lack of public awareness is certainly a contributing factor. My hope is this book will help remedy the situation.

Next is a story about a bold attempt to jump-start the acceptance of laser machines by a skeptical industrial community.

---

[96] The Gemological Institute of America, GIA, is a reliable grader of diamonds. They grade only, appraisals of monetary value are done by other companies, a list of which is on the GIA.edu website under About Us then Library or use the search box on the home page.

[97] A History of Diamond Treatments, Overton and Shigley, Gems & Gemology, vol. 44, (1), 32-55.

WITH KORAD'S
K-Y2 SERIES

GUARANTEED LAMP LIFE

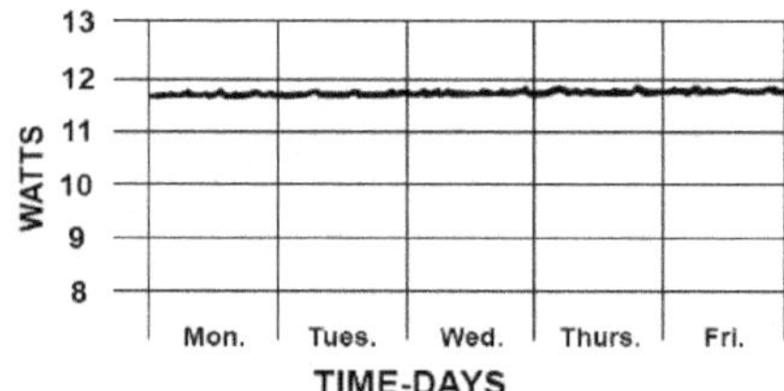

This versatile laser provides output with long-term stability, in three distinct configurations:

- •CW multi-mode power to 12 watts
- •CW TEM_{00} power to 1.5 watts
- •Q-switched power to 10 KW, and rep rates to 1200 pps

Capabilities of the K-Y2 extend to frequency doubling, mode locking and acousto-optical modulation.

For further information on how the K-Y2 can fill your needs, call (213) 393-6737, Ext. 292, or write Korad's Sales Manager

UNION CARBIDE IS A TOTAL SUPPLIER TO THE ELECTRONICS COMMUNITY OFFERING THE FINEST IN CERAMIC, FILM-FOIL AND SOLID TANTALUM CAPACITORS, CRYSTAL PRODUCTS, LASER SYSTEMS, AND WATER MONITOR INSTRUMENTS.

Union Carbide Corporation
KORAD Department

KORAD® Department 2520 Colorado Avenue
Santa Monica, California 90406

Korad's gift of technology to the world thanks to Dr. Koechner, Bud Erickson and Lee Benson. From 1968-69 Laser Focus ads.

Chapter Eleven: The Korad Laser Van

Korad's Laser Van, 1969
With Jim Kessey and his Mother in Detroit

Laser Circus Hits the Road

The time was late 1968 when I first heard of the "Van Project," as it was mysteriously called, from my boss Bill Thurber, who told me I was to be in charge of it. He was the marketing manager replacing Tony Johnson, who had quit in June 1968 after Maiman had left the previous December. Indeed, the entire top management team bugged out after Maiman took off, including Burns, Lubin, and Hoskins. But we who remained did not worry about the vacuum of leadership; we knew lasers quite well indeed. And Carbide's management replacements,—Dr. Clayton (Clay) Zerby, and especially Bill Thurber—were more than capable of running the company. And we Koraders were proud to be part of a giant, well known company.

Unlike everybody else who had occupied the marketing manager slot, Thurber was a first-rate engineer and budding businessman who was to become like Ed Young—a department president within Carbide. Unlike Zerby, he completely understood our laser technology, even authoring a paper on laser welding. He was to replace Zerby as head of Korad in mid-1971.[98] Zerby just couldn't manage to get his arms around our laser technology. I think he was too theoretical and not enough of an engineer. I tried hard to educate him but he complained about my sending him too much detailed information in the company mail.

The idea behind the laser van was to put working lasers into a standard mobile home and drive it around the United States visiting companies, government labs, colleges, and universities. Lasers were commonly said to be a solution in search of a problem, for these were the days before the explosion of applications for laser technology had occurred. Had we only known how widely accepted our lasers would come to be! We were to bring working high-powered lasers to the customers and, in doing so, prove "robustness" as well as performance.

Robustness was big with Thurber. After all, if we could go clattering and thumping, bumpity, bump, bump over zillions of railroad tracks, that would surely attest to the laser's robustness, so the reasoning went. All these miles were to be logged behind the wheel of a brand-new, 1968 thirty-foot-long Dodge (and Travco[99]) motor home. This thing was something for its time—not a square corner on her. She looked like an Oscar Mayer wiener on wheels (minus the bun). They were really going to pay me and a couple other guys to run this impressive old girl around the country! Motor home it was, but not for us, we slept in motels. Success would proclaim robust, hardy lasers

---

[98] "Thurber Promoted," *Laser Focus* (June 1971): 48.

[99] Dodge motor company built the chassis; Travco, located in Brown City, Michigan, modified it into a motor home.

had arrived and were available from Korad whose laser solutions worked and were ready to solve problems.

Korad may not have been the least expensive laser supplier in those days, but we had the advantage of our association with Maiman, the man behind the technology. The scientists who would employ these devices for their experiments knew that name, and that fact, along with Union Carbide's backing, helped position Korad ahead of all other laser suppliers. Besides, Korad's lasers were known for simplicity of design and reliability. Having a Korad laser conferred pride of ownership in the scientific community. After Maiman left, I became the sales manager in charge of industrial laser machines. I vigorously promoted the Carbide name and image to make up for the loss of Maiman. I was exceedingly successful, and sales were booming.

The industrial community was finally coming up to speed, but where lasers were concerned, industry was still in the dark. We used to say, sometimes with some frustration, that *all* Americans must come from Missouri, the Show Me State. This was especially true of our engineer customers. Therefore, we put laser machines, a trimmer, and a welder/driller on the van. However, there was no laser marker of identification numbers and barcodes, as those computer-controlled devices came about ten years later, in the mid-1970s. These were the early days of the laser era, the decade of the sixties. Sometimes it had its comical aspects as we wheeled through various appointments in one city after another, our straight-laced business attire and demeanor contrasting with the counter-culture attitudes, beads, and granny dresses of the time.

But regardless of their backgrounds, everywhere we went everyone turned out to see a *Death Ray* in the van. It must have been a little disappointing to them in retrospect, for H. G. Wells we weren't—no scorching of the Earth, no maiming of innocent civilians or livestock, just a couple of middle class working stiffs trying to make ends meet.

Until the mid-1970s, none of Korad's machines had digital computers to control and manipulate the laser beam. But we did have an analog computer to control our semi-automated laser resistor[100]

---

[100] Resistors resist the flow of electricity. Usually they are printed with a resistive ink on thin ceramic sheets called substrates. The ink slops down rather uncontrollably. So they print 20 percent more ink and then trim off the excess with a laser by vaporizing the baked on, now-hard ink. Accurate resistors make accurate electronic

trimmer that was introduced after the van project ended. Digital computers in those days were incredibly primitive compared to modern ones. The gold standard was a PDP 11 minicomputer made by DEC, Digital Equipment Corp., which cost $20,000—an awful lot of money. Besides, computers were hopelessly slow and just not worth the money and hassle. Virtually all laser machines of that era were manual. None had robotics or machine vision, and we certainly didn't even think of air scrubbers to remove smelly odors or even carcinogenic vapors. Union Carbide, now 100 percent owner, insisted that the Carbide logo be prominently displayed on both sides of the van.

In December 1968 the van was picked up in Brown City, Michigan and driven it to Korad in Santa Monica. By March 1, 1969, I had it fully outfitted and ready to hit the road. It was to serve as a "travelling billboard" promoting the Union Carbide name throughout the United States. This billboard concept was so valuable and enticing to Carbide that they paid for all thirty feet of that monster Dodge, an offer Korad couldn't complain about. To Carbide's top management's way of thinking, this billboard concept justified the cost. However, we weren't allowed to hit the road until the logo was perfectly acceptable to the top brass of the mighty Union Carbide Corp. The van had to be repainted a few times before we got it right.

Do You, Rod Waters, Take This Thirty-Foot Dodge

I was directed to first study the layout of the standard thirty-foot long Dodge/Travco mobile home, figure out how to stuff as much laser equipment into it as possible, and then be the engineer in charge of outfitting it with two roof mounted five-kilowatt generators, a refrigerated laser water cooler, and sundry specifics. Then finally, when the show hit the road, I was to be in charge of scheduling where and when the demos were to be performed—and, by the way, also occasionally conducting demos on the van as a backup to the primary salesman, George Kendall, who would do the usual wheeling and dealing while my other counterpart in sales, Jim Kessey, was to be a second backup to be involved in demos as my counterpart in sales. He was in charge of scientific product sales whereas I pursued clients in industry. Two of the three lasers aboard were industrial, so it was to

---

devices for everything from heart pacers to car radios.

be "my baby." At least that's how I looked at it; my thirty-foot-long gas guzzler forced upon me, travelling billboard indeed!

This was an important gig because industrial laser machines were Carbide's focus for Korad's future, and indeed, industrial products were to eventually become Korad's main line of business. There's a simple reason for this: scientific lasers were "onesy, twosy" as they say in sales, whereas industrial were, again in the language of sales, "cookie cutters." You could sell industrial in volume and make a very nice profit, and you didn't need scientists to both build and sell them. Street people do nicely—well, *talented* ones do.

As an aside, I know this from my own experience of starting Florod Lasers in 1974. I ran it with my partner Floyd Pothoven until 1996 when we liquidated it and I created Laserod, in Torrance, California, out of it. I ran and owned Laserod until 2011 when I sold it at age 75 to become their marketing manager. Both companies, Florod and Laserod, were into exclusively industrial lasers, a field I have been in for almost a half century with not a day lost to unemployment. The Korad "laser van" was to be instrumental in my success at Florod, where I used the concept of traveling around the country selling a laser machine carried in the back of my Chevrolet station wagon. It was hardly a thirty-foot Dodge, to be sure, but I never had pockets as deep as Union Carbide.

Want a Snort?

George Kendall, as mentioned, was to assist the sales team as Korad's eastern regional sales manager. George was a really nice guy, jovial and rotund, and with an abundant thatch of sandy hair immaculately barbered. He seemed to operate with a full head of steam, routinely fueled by Dewar's White Label Blended Scotch Whiskey. Tom Bionde, the designated driver, remembers having a grand old time with Kendall at a San Jose strip club. But then, most of us had an interesting story or two where George was concerned. My dominant memory of him over the years involved one particular night in his hotel room after dinner, at which time he challenged me to a drinking match. The last man standing would win, he advised me. He was an ex-Marine, a big man, and renowned as a prodigious drinker. While recalling any number of lunches with him and clients during which he downed three strong drinks while remaining seemingly cold sober, I declined. My prospects for victory in such a contest seemed rather bleak under those circumstances.

Bionde had been an electronics tech at Korad before being promoted to drive the laser van and to be in charge of operating and maintaining the laser equipment. This consisted of three lasers: a ruby laser Model K-1 in the back, a YAG laser resistor trimming machine on the right side, and a ruby laser welder/driller on the left side just behind the driver. It even sported a holography exhibit in the back next to the scientific ruby laser. The hologram showed 3D images of parts such as a triangle, a ball, and a pole that you could look around and see behind as you moved your head side-to-side. It was a simple holographic plate illuminated by the van's lighting. This was the first hologram that virtually all of the van's visitors had ever seen.[101]

I remember Kendall as a hefty man, imposing, super friendly, and a natural salesman. When Bionde took his week off to recuperate at home in California, he simply parked the van at his departure airport. The company would fly him back and forth from van to home and back. Tom remembers me as his only boss whose orders he followed because "that's my nature." He used to kid me about my name, saying I was Sky Blue Waters and sometimes Muddy Waters after the famous blues singer.

Come One, Come All

Bionde was really in his element as he piloted the motor home across the countryside. One afternoon might find him and the crew chatting up industry folk at Texas Instruments in Dallas, Texas; another afternoon we might be hawking our wares while sharing a cold beer and jambalaya to the strains of some good blues in New Orleans. Next thing, we'd be farther north in St. Louis or over to New England where we hit many colleges and universities where many got their first look at lasers. We did New York State and all the IBM factories along New York's Hudson River Valley, then Long Island, and New Jersey. Bionde was having the adventure of his life! Of course our home state of California did not escape our attention, nor did the Northwest. All the way up the north-south I-5 corridor connecting Mexico to Canada we

---

[101] Holography or 3D pictures, although invented in the 1950s, required the invention of the laser to become a reality.

went, to Santa Cruz, San Francisco, and even up to Washington State and the busy Seattle area.

The laser van was immensely popular and thousands took the tour. At every stop, people clamored to see it—even the janitors! I remember trying to control who visited without much success. People begged for admittance. Nobody had ever seen a laser, but everybody had heard of the "Death Ray" and wanted to see it for themselves. Tom related how he would shoot holes in pennies for visitors to take away as souvenirs. The ruby laser on the right side did the drilling as well as the welding.

A typical visit to a possible customer company or college went like this: We would wheel into the parking lot, sometimes containing thousands of cars, and select a spot close to the main lobby. Tom would immediately level the van on jacks so that it would not pitch about as the stream of curious filed through. We had two doors, an entrance on the right front and an exit at the right rear. Bionde ran the lasers while Kendall, Kessey, or I did the selling.

As eye protection for the spectators, Bionde built what he called a "blue coffin" around the powerful hundred-million watt scientific laser in the back. It was made of Plexiglas designed to block the laser rays. We were very mindful, even afraid, of high-power lasers. After all, the "R" in laser means radiation. Another form of radiation, radioactivity, discovered by Henri Bequerel in 1896, rewarded some of its early pioneers with death. Determined not to follow them to an early grave or to blindness, we took precautions such as the blue coffin.

The laser van was a trouper, and we definitely broke the old girl in. After we had piloted her over every mountain, desert, and valley from sea to shining sea, the repair bills were becoming relatively frequent while sufficient sales were not being made. The van was mothballed and probably ended up painted in sunflowers and peace symbols somewhere in San Francisco's Haight-Ashbury, but it had toured the United States visiting 35 states for two years, racking up over 50,000 miles.

Korad's Only Episode of Laser Blinding

We had an actual instance of laser-beam blindness when engineer Earl McLendon suffered a high-power ruby laser beam directly into his eye. McLendon's boss, Walker, took him to UCLA's hospital where they had world-renowned eye doctors. There was nothing the doctors could do for him. Fortunately for him, after some months went by, he recovered nearly full use of his eye. We used to have a consoling joke about this: at least you could only lose one eye at a time!

Kendall's employment was terminated after he hit a support cable on a telephone pole with the company car, not the van. Lucky for him the car went up the cable about the height of a big man. He was let go by Thurber who asked me for my opinion on the matter. I dryly observed that Kendall's fondness for Dewar's would doubtless lead to more accidents. Kendall then went on to work for Lytron, selling YAG crystals.

The Korad Laser Van Helped Establish Laser Machines

Laser welding and drilling did not exactly soar into the skies and explode, resulting in a shower of orders; sales of these machines remained at five per year. But the resistor trimmer took off and sold by the hundreds. The basic problem with printing resistors, similar to a printing press, was they used ink that could not print accurate parts. Then the electronic circuit, for example a car radio, would not work.

Laser trimming made fast and automated production of automobile radios possible. Starting with the 1970 model year, General Motors Delco Radio Division in Kokomo, Indiana, began producing all car radios for GM using this new process. Within a few years all car radios in the world were laser trimmed. The 1970 GM Delco laser was a gas CO2 type. In just a couple years I had convinced the industry to convert to the solid-state Q-switched YAG laser, which remains to this day the gold standard.

So what else are laser trimmers good for? How about music on your computer or DVD? The core devices making this possible are called analog to digital converters (ADC). These convert music to be digitally recorded into digital numbers, zeros and ones. Then to play back the music, DVD or whatever, you use a DAC, digital to analog converter. Why use lasers to trim the devices? -- to get accurate numbers, for example if you wanted to digitize music in the form of a voltage (wave), you would do it on some scale like zero to ten thousand with zero representing zero volts and ten thousand, say ten volts. So to get hi fi music, you need the smaller numbers (millivolts, say) to be accurate. The best way, then and now, is to laser trim the printed circuit to an accurate value of resistance. Now you can enjoy your music.

Laser trimmers replaced abrasive trimmers that used an air-driven blast of sand to remove or trim small amounts of resistor in order to get an accurate resistance value. The trouble with the sand blasters, as we laser guys called them, besides being messy, was they clogged and wore out any nearby moving parts. The nearest moving thing was usually themselves, so they self-destructed. They were just too dirty and slow, so they had to go. Besides, they weren't accurate enough.

I certainly played my part in championing the switch to YAG laser trimmers while promoting the latter technology during the Carbide van tour. I had touted YAG tirelessly as I flew about the country attending all the ISHM (International Society of Hybrid Microelectronics) trade shows that took place every couple months throughout the year. The van had so much potential when utilized correctly that I adopted this strategy years later in my own business at the helm of my far more modest promotional vehicle, my station wagon. And while it was quite effective as a strategy, it still never lived up to the throaty growl of that old Dodge as we climbed yet another mountain grade, or her steady hum as we floated across the Midwest at sunset.

How Korad Changed a Life

Bionde didn't miss his old life back at the factory; on the contrary, he was very pleased with his exciting van job as he went instantly from a technician to an engineer without a degree in engineering! Discussing this period with Tom was similar to many interviews I conducted with other Koraders in preparation for this book. Like them, he enjoyed his time at Korad and felt he had learned a lot. It changed his life because he decided not to ever work for anybody anymore. And the decision came not from any bad experiences, but just from realizing he had the ability to work on his own without supervision. Korad was a place where talented and dynamic individuals were given much more latitude; a place where their input, personal experiences and aptitudes were genuinely respected and nurtured.

After the van was retired, Bionde was promoted to field service engineer and continued travelling around the country troubleshooting and repairing lasers. He did this for a couple years. It was his vast exposure to life in the United States that brought him to the decision to work for himself for the rest of his life. He sold insurance for a while, and then had a swimming pool service business, which, he said, was "very lucrative." He owned a surf shop for the last ten years before retirement on Hawaii's North Shore. The shop made enough money to buy him and his wife, Jeannine, a nice home on an acre of land on the North Shore. But the store was too much work, so they gave it up. At age 81 he still surfs and plays tennis. He says he is "having a great time being retired." Wife Jeannine is from France, is a youthful age 76, and still works as a tax preparer.

Chapter Twelve: Parasites and Patent Law

Among Korad's many woes from the mid-1970s on was the outcome of the patent war that had raged for over ten years between the laser industry in general and Gordon Gould, an optical and microwave spectroscopy student under Nobel Laureate Dr. Polykarp Kusch at Columbia University. Gould had been pursuing his doctorate in that field during the late forties and fifties, and eventually found himself sharing his ideas regarding the MASER with Professor Charles Townes, also of Columbia, and an eventual Nobel Prize winner.

The Gould Patents Further Sap Korad

While Ted Maiman undeniably built and fired the first laser device, we can perhaps give Gould credit for coming up with the acronym LASER, light amplification by stimulated emission of radiation, which he popularized as the focus began to shift from amplifying microwaves and toward amplification of light. Bell Labs and Townes were pushing hard to name the laser an "optical maser."

As the race to build the first laser got underway Gould was in the thick of the discussions and developments at Columbia, but his primary focus and methods were the subjects of considerable concern and frequent scorn among his colleagues. Townes and others found Gould to be an erratic student, one whose focus throughout the race, as detailed by Jeff Hecht in BEAM, was first and foremost financial enrichment for himself. While everyone else was preoccupied with discovery for its own sake, Gould plotted and schemed to secure for himself credit for innovations in lasers. The subject of patents came up in his first discussions with Townes in 1957 and seemed to singularly animate Gould for the next couple decades.

His approach, as he maneuvered to fine tune his patent applications, involved the careful compilation of notes and drawings in a formal bound notebook, intended to demonstrate what he knew and when he knew it. Gould had his notebook notarized at a local candy store in 1957. The very next year saw him approaching a New York patent law firm, Darby and Darby, recommended by Gould's parents. It was there that a significant misunderstanding took place. Gould came away believing he had to produce an actual working model in order to get a patent. It had been unsuccessfully explained to him that he needed only to produce a sufficiently detailed set of instructions so as to allow one skilled in the field to build one. Since that time it has been the source of much speculation whether Gould would have been granted a patent had he not labored under that misconception and simply gone forward with his application.

Gould joined Technical Research Group and began work on a more substantial notebook. TRG had been founded in 1954 and was the happy recipient of piles of Pentagon research funds. Ostensibly working to develop atomic clocks that produced a precise frequency that could serve as a time standard, Gould was fervently preoccupied with his laser-related machinations. He was even able to circumvent TRG's standard patent rights agreement, typical throughout industry, in which the company reserves for itself ownership of any patents that may emerge from the work of its employees.

Gould subsequently was able to persuade TRG's physicists that bouncing a beam of light back and forth in an optical resonator could potentially produce tremendous power, an idea altogether compelling to the Pentagon. But the Pentagon brass were conservative folk who didn't care for risk and, as a result, had just lost the initiative to the Soviets when Sputnik's pesky beep, beep, beep embarrassed and alarmed both the American scientific community and the American public.

To avoid another such distasteful incident, the Pentagon inaugurated the Advanced Research Projects Agency, specifically chartered to pursue and exploit daring, out-of-the-box ideas, like producing a coherent beam of intensely focused light energy. Gould was quite a salesman, and ARPA was impressed, contracting with TRG to proceed with development of the concept. The problem was that the project required the intense cooperation of Gould, who was precluded from participating directly because of his lack of an appropriate security clearance. A youthful flirtation with the Communist Party and his subsequent refusal to identify his fellow sympathizers had stood in the way of a top secret clearance, which ARPA work required. Basically, the reason that Mr. Gould suffered for nearly thirty years getting his laser patents was his membership in the Communist Party, the enemy of the Free World during the Cold War which lasted nearly a half century.

Commie Ties

Gould's security issues negatively affected his fervent quest for laser patents. The trademark office cannot issue or publish material that is classified, and, of course, with no patent there are no royalties. Resourceful as ever, Gould's lawyers succeeded in splitting the patent application into two parts, the shorter of which they expected would evade classification. During this clever maneuvering, those around Gould became increasingly suspicious as he seemed to be a virtual spigot of ideas, gushing forth with one brilliant idea after another. Q-switching and the transfer of energy between colliding gas atoms were two of Gould's most recent ideas when Townes had a look at TRG's proposal to the Air Force.

Jeff Hecht discusses Towne's reaction in *Beam*.[102] "He didn't think an indifferent student like Gould could have come up with all the ideas by himself." Townes was particularly surprised to see that the ARPA proposal also covered transferring energy from colliding gas atoms. That was Ali Javan's idea, and Javan had not yet discussed it openly.

You can bet, however, that he had discussed the possibility at some point with Gould, whose little notebook was swelling with ideas by the month.

---

[102] *Beam: The Race to Make the Laser*, published 2005 by Oxford University Press.

Acle Hicks, who had a master's in physics and who worked for Korad Lasers from 1966 to 1968, had a particularly distasteful reaction to Gould, which he shared with this author in an interview at his Cupertino, California, home in March of 2010. Hicks was one of the Boyden Group who left Korad to work for Holobeam, a Korad copycat laser company in Paramus, New Jersey. Gould was heavily involved with Holobeam on a regular basis.

Hicks said, "I met Gordon Gould a number of times and I can tell you a few stories. I was a new engineer, right out of school, and he would come around and brag about how great he was. One time he made the rounds with a separate lab notebook for everybody—a nice, shiny new scientific notebook. He gave one to me and he said, 'Look, every idea you think of, I want you to write it down in this notebook. Keep it carefully documented. I'll come around every month or so and go over it with you. You sign and I'll countersign. If there's something patentable in there, we'll patent that together."

Another laser engineer, who wishes to remain anonymous, told me that Gould was a certified lowlife, visiting the other engineers and requesting them to contribute to his hoard of unethically gained patents. Gould was busily grabbing credit for other's patentable ideas. "Sleezy behavior," my other source remarked and went on to say, "He talked about his yacht and how he screwed the government on taxes. He would pretend to rent his yacht to people to make it into a business so he could save on taxes." He finished by saying, "Every interaction I ever had with the guy, I felt like washing my hands afterwards."

Another of my Korad interviewees, Carl Schulthess, related how his physics professor at Cal State Fullerton, told him about being a fellow student of Gould at Columbia University. This professor, now deceased, claimed that Gould regularly fed on others ideas and raced off to patent them.

As Gould continued his attempts to endear himself to his various colleagues, his cadre of lawyers was hard at work, trying this approach or that, which might afford them success in securing a patent or series of patents related to the laser. Throughout the late sixties and early seventies, they had sought patents on the laser itself, but were unable to convince the patent office to play along. Why not simply attempt to patent one or more individual components of the laser, rather than the instrument in its entirety?

For example, he had the idea of patenting optical amplification at the atomic level. This clever idea went to the meaning of laser: light amplification by stimulated emission of radiation. Totally brilliant to patent that! His shrewd strategy was ultimately successful, as his career-long thirst for personal enrichment was quenched in 1977 when the first of a series of patents on various laser components was issued. He and his backers in 1979 formed Patlex, a company responsible for holding the patent rights and handling issues related to licensing and enforcement. Thus Gould and his lawyers became millionaires, but this boon to them was only realized by imposing huge costs upon the industry,[103] costs which the ailing Korad was ill-equipped to cope with.

On the other hand…

Every coin has two sides; here is an opposite side of Gould. Dr. Don Sims, fourth president of Korad, told me that he and Gould had worked together at TRG. He was a "good" friend and even invited him to sail on his boat a few times. Sims described him as very intelligent, creative, and nice. Gould was denied a security clearance because he had attended communist meetings with his girlfriend. As the government-funded laser research on TRG's Project Defender was classified, Gould was prevented from working on this million-dollar contract to develop the first working laser. Sims felt it was self-defeating for the government to do this "because they kept a very valuable intellect from working on programs that he should have been allowed to work on."

However, it was still the era of anti-communist "witch hunts," called McCarthyism for the senator from Wisconsin, Joe McCarthy, who had conducted the most notorious witch hunts. And it was less than ten years after the Rosenbergs had been executed for giving the Soviets atomic bomb secrets. The Pentagon was not about to allow this fox in the Death Ray chicken coop!

For a long time I have wondered if Gould was a political animal, so here was my golden opportunity to find out. I asked Sims, "Did he ever mouth off about communist causes?"

The reply was, "No. Absolutely not. Absolutely not."

---

[103] The Gould patent royalties were as high as five percent, far higher than the typical percent or two. Consequently, they were hated by all those companies who suffered under their predatory rip off of the laser industry, me included.

As to whether Gould had stolen laser ideas from Professor Townes at Columbia University where Gould was working on his doctorate in physics, here is Dr. Sims' view: "He was a very intelligent guy, very creative. The famous patent story—which I know you must have heard because it's a legendary story—is that he was working as a graduate student in the Bronx at Columbia, and Charles Townes, of course, was the head of the physics department. Gould was working on some laser-type concepts. This was after the publication of the Arthur Schawlow-Charles Townes paper, the famous one of 1958. Schawlow was Townes's brother-in-law, so they had a good relationship."

However, Gould was working not for Townes but for Columbia Professor Polykarp Kusch, Sims explained. "Kusch wouldn't let him work on a laser-related project because the probability of success was too low. He wanted students to have a defined path where they would work on their project, write a thesis, collect data, and get their doctorate degree."

Sims continued, "Gould was not part of the group of Columbia University graduate students who were working with Professor Townes on the laser. What happened is he had some ideas that were very, very clever—patentable, in fact, so Gould got worried that somebody was going to preempt him, publish first or patent first—and you know that's death both in the world of academia and of business."

Laser author Jeff Hecht has some good things[104] to say about Gould: "He was the first to think of lasers as high-power devices and first to think of laser-beam weapons such as target illuminators, laser radar, and secure communications."

Next we explore a few of Korad's more interesting laser applications from Hollywood to holography to medicine.

---

[104] Personal communication.

Chapter Thirteen: A Diversity of Laser Applications

Donald (Don) Smart[105]

Well, these years were not all hijinks and highballs at Korad, for the laser's potential had been discovered in a big way. Industry had come to realize that this fledgling technology could potentially provide the answer to a myriad of difficulties, and not just in the realm of science. Hollywood even sat up and took notice.

---

[105] Don Smart has two bachelor's degrees in chemical engineering and electronics engineering, a master's in systems engineering, and lots of classes in physics at UCLA Extension.

Korad Goes to Hollywood

Korad landed an opportunity in the summer of 1970 to provide a laser for *The Andromeda Strain*[106] movie in which it bedeviled the movie's protagonists as they struggled to contain a dangerous contagious virus in a quarantined, secret government facility. The movie was nominated for two Academy Awards. The film featured a Korad K-Y2 laser whose job it was to kill escaped mice in a laboratory. The director, Robert Wise, wanted the laser in three days. Korad had it to him in less than a week. Korad accepted his offer of $25,000 for the laser rental including support by Koraders Len Debenedictis, manager, Dr. Dennis Rice, physicist, and technicians Craig Zahnow and Danny Floyd.

The project was initiated on a Friday morning; the laser was delivered late Sunday evening; and filming began the next day, Monday. Korad personnel were present as needed during the next two weeks of filming. The studio was a big windowless warehouse structure located on Universal's main complex, now a major tourist attraction. In the basement, actually a pit, they had constructed a proper scientific laboratory called a clean room. It was a spotless white and quite impressive.

Director Wise was definitely the man in charge. People jumped to do his bidding; he got what he wanted. Interestingly, he did not use a standard canvas director's seat. Rather, he had a big living room easy chair which was lowered into the pit so he could direct the action in comfort.

Dennis Rice, still occasionally working in his position as Assistant Dean of Engineering at the University of California at Riverside, said the Korad K-Y2 was the backbone of the system he assembled for the filming at Universal Studios in Hollywood. He described it as a Q-switched continuous wave neodymium doped YAG that was frequency doubled to the green. At that time only mechanical Q-switches were available for continuous beams. This one was a vibrating rear mirror producing about a thousand laser pulses per second, fast enough for the beam to appear continuous in the film.

Watching a movie being made was definitely an out-of-the ordinary experience for Rice. He reported it was most interesting seeing the film

---

[106] The *Andromeda Strain* is a 1971 American science-fiction film based on a novel by Michael Crichton. The laser was a green YAG to make the beam visible.

being made in pieces, later to be spliced together in the cutting room. The Korad laser, however, was just one scene in the movie, taking two weeks to produce. There were lots of "takes".

Most remarkable was the large number of people involved, hundreds of them. Some were standing around watching but most were working away at a steady pace, not at all hyper. Hollywood is dominated by labor unions, consequently most workers were single task types. The exception was the "grips" who were jacks of all trades. Their grip was their main go-to guy. Movie making is expensive, time is money, so there was considerable pressure to keep the laser running. People seemed to enjoy their jobs and felt they were "special". After all, who doesn't have an interest in movies and their making?

Physicist Rice recalled one particular difficulty: getting the laser beam to appear crisp and readily visible on camera. A fine mist of oil was sprayed in the air, which seemed to achieve the desired effect, but the oil mist blocked the beam and stopped the laser from working. It seems the mist had deposited itself on the laser mirrors which temporarily killed lasing. His helpers merely installed inexpensive plastic bellows tubing between each part of the laser, the head, the Q-switch and the mirrors to keep out the mist. A valuable lesson was learned: forever after that all Korad's industrial lasers were sealed with bellows-like small accordions. And to make doubly sure of success, he had a nitrogen purge installed that flooded the laser with gas, thus blocking the oil mist.

As Dr. Rice explained, "Part of the agreement with Universal Studios was that the laser would not injure anyone. However, Universal ignored this and in one scene the intense green beam was directed at an actor's face. Then the actor's cheek was covered with green make-up to show how the laser had damaged his skin. Fortunately, they did not direct the laser at the actor's eyes. The beam consisted of twenty-five thousand watt pulses which could result in significant eye damage."

Carbide leadership was proud to see its name in the film credits. To this day, should you view it, you will see Carbide's name and logo in the credits. The movie lives on, but Carbide is no more. They sold out to DuPont Chemical to escape the liabilities stemming from the Bhopal, India, disaster on the night of December 2-3, 1984, when many thousands of people were killed and injured, some permanently, as a result of Carbide's chemical plant leaking poisonous gas.

From Moving Pictures to 3D Pictures

Smart also discussed some amazing work involving holography. Invented by Dr. Dennis Gabor in the late 1940s, it took lasers to make the concept work. Ordinary 2D photography gets only the amplitude of the light wave; 3D holography gets both amplitude and phase, which makes the image three-dimensional. Laser light illuminates the subject, and the results are recorded on photographic film. Playing it back with light or another laser reproduces the 3D image. The film can be cut up and will still produce a complete image, but it has lower brightness. Interestingly enough, the human brain shares this holographic feature: pieces of the brain can be removed yet still retain memories.

Korad's contribution to holographic photography was a holographic camera that enabled photographic analysis of moving components exposed by flashes of laser light mere millionths of a second apart.

Korad was an innovator in double-pulsed holography for nondestructive testing of things that move slightly toward or away from the holographic camera. It was capable of measuring parts that moved (displacements) down to a third of a micron—the wavelength of ultraviolet light. But it was a niche market lacking big profit.

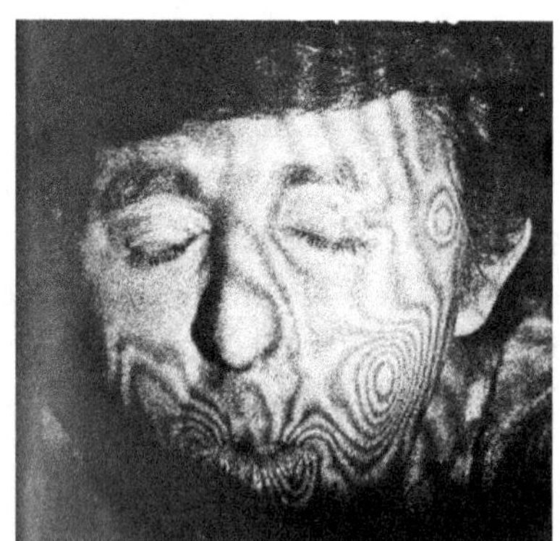

Double Pulse Hologram of Ben Parks, Electroniker

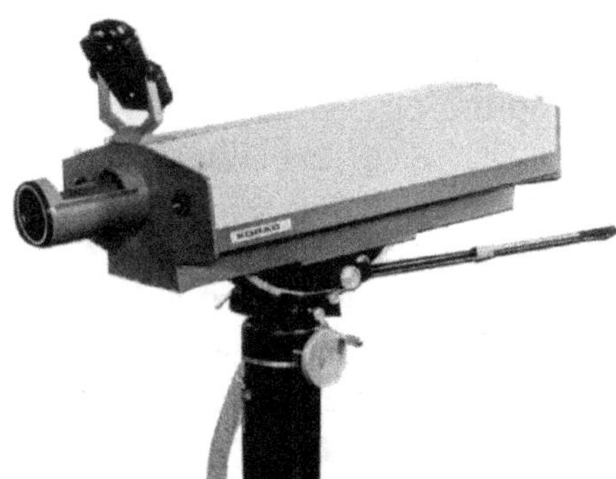

Korad's Holocamera

Don Smart again: "Oh yeah; I loved that! I built the holographic camera that was delivered to Mitsubishi for doing turbine blades for their supertankers." Just as cold temperatures led to the *Challenger* space shuttle tragedy, so too did December temperatures pose problems for Mitsubishi.

He continued, "I spent six weeks in Nagasaki on that one; initially it didn't work. They wouldn't pay for it unless it worked. That's why we stayed there six weeks. The big problem was it was December in Nagasaki. They had it installed out in the open in a big, cold, drafty warehouse."

The temperature was changing with the drafts causing the length between the laser cavity mirrors to change, thus changing the laser's wavelength (color). Holographic lasers must operate with very pure color beams in order to get good holograms (3D pictures).

Mitsubishi technicians followed Smart's and his colleagues' advice, and moved the system into a heated room. The problem was solved and Korad got paid. It was fascinating stuff, this type of photography. As Smart explained, "It could allow for the detection of impossibly small vibrations in a turbine engine by using a double-pulsed laser to take two pictures millionths of a second apart, and one could see what had changed in those millionths of a second. Each of those close-together laser pulses, millionths of a second apart, were individually *billionths* of a second in exposure time. It was a very fast-freeze photograph."

In other words, if you take a hologram while something changes, and then you take another right on top of the first but only millionths of a second apart, you would see any shift, either horizontal or vertical. Smart added, "Korad sold a fair number of those, maybe as many as fifty lasers for holography over those few years. Not a bad project."

So, How's Your Howitzer?

The United States Army, as one might imagine, found such research into photographic holography compelling too, as in its efforts to detect weak points in 105-millimeter howitzer shells. Personnel at an armory in Dover, New Jersey, were being periodically injured when a shell would explode in the gun. Smart recalled how they turned to holography to find a solution to that problem.

"They'd take a hologram of the shell and then pressurize it with air. If it had a defect in it, it expanded differently where the defect was. It made kind of a jiggle in the hologram. They did this on every single shell that they made."

Listening to Don Smart going into detail about these applications made me recall my life's work in industrial laser applications. I wish I could have been able to gather all the people I had known in that very room while I interviewed him so that I could watch their reactions, his gestures being more animated than ever. His example this time dealt with particle sizing associated with Boeing jet engine nozzles.

Smart continued, "Yeah, that was interesting. It was a single-pulse hologram. A nozzle shoots out a series of particles, and you took a hologram. *Bang!* The hologram would freeze the particles in a three dimensional space within the two dimensional hologram. Then when you viewed the hologram, you could focus through the particles and measure their size and distribution." Then Millipore engineers could design a better engine.

Medical Research

The first half of the seventies also saw research into glaucoma, where attempts were made to relieve eye pressure by drilling a tiny hole in the eye. The laser beam was delivered down a fiber bundle. Now this does not sound like the stuff of amusing anecdotes, but Smart was quick to share an interesting story involving a particularly unfortunate Rhesus monkey. The National Institute of Health (NIH) bought a Korad laser, and they practiced on live monkey eyeballs. One day they phoned Smart and said, "Don, the laser didn't close down when the power went too high, and it ruined one perfectly good Rhesus monkey. The laser didn't fire the first pulse; it fired a second time, and it blew the eyeball all out."

Smart felt terrible about this. He went to his California laser lab and found if the laser didn't fire correctly, the power went twice as high. Then he was able to develop a solution. He travelled cross-country to the NIH and went down to their basement where he discovered his customers were a couple of colonels in the Air Force, medical doctors in Hawaiian shirts, and they had their feet on the desk. They put an eyeball under the laser and drilled a hole in it. Under the microscope, they showed him how the hole looked. "They had a whole Petri dish full of eyeballs."

He clumsily knocked the Petri dish of eyeballs onto the floor where they scattered about. Embarrassed, Smart went down on his hands and knees to start collecting eyeballs.

The two doctors assured Don they were not concerned and quickly began gathering up the specimens. No doubt relishing the opportunity to make sport of a newcomer, one of them cavalierly tossed an eyeball in Smart's direction. "I grabbed this bloody eyeball and said, '*Oh my God!*' They thought that was the funniest thing that ever happened."

Port-Wine Blemish and Tattoo Removal

Smart was a spigot gushing at full volume where laser applications stories were concerned, and I found myself virtually on the edge of my seat as he continued with his recollections. While the eyeball-drilling idea never met with commercial success, apparently the thawing of the Cold War had Smart contemplating a different type of laser procedure on then-Soviet Premier Mikhail Gorbachev. "We sold a number of these machines for removing port-wine stain blemishes on human bodies, typically the face. That was a good project. Should have operated on Premier Gorbachev of the Soviet Union; he had a nice port-wine birthmark on his face."

In another instance, one of Korad's clients, a plastic surgeon, had removed a similar blemish from a patient's chest using a Korad laser. He was so impressed with the results that he came to Korad with photos of the procedure in hand.

"He showed us these pictures. And it was just a brute force like five- or six-joule one-centimeter beam. He had just strapped the poor patient on the end of the laser, and it was bang! Bang! Bang! Just as crude as anything. It was a red ruby laser, and he showed us pictures right after; and the person's skin, it just removed all. It was raw, just awful. I asked the surgeon if that hurt, and he said, 'Well, we give them a little sedation. It takes a tenth of a second for pain to get from here to the brain, and by that time it's all over.' Six weeks afterward they showed us the patient again, and you couldn't believe it. The skin was just perfect." Unfortunately, such lasers never were deployed to the American market as they did not at that time have FDA approval, a lengthy and costly process. Thus Japan, which required no such approval, became the primary purchaser for such units.

Related to blemish removal is tattoo removal by laser, now a big industry. Those who are in this business can find very professional looking equipment on the Internet. There are three main lasers used for zapping tattoos: ruby at 694nm wavelength (red beam), YAG at 1064nm (infrared invisible beam) and YAG at 532nm (green beam).

Each laser is used for certain colors only. For example ruby is good for erasing green, blue and black inks, the other lasers remove the rest of the inks.

The Biggest Gorilla on the Block

The year was 1967. By far the most frightening and scary laser I ever worked on was one so powerful that nothing could stop it. This CO2 laser beam would burn through everything it contacted including people, whom it could maim or kill. I wasn't terrified of it; I just didn't want to spend much time on this monster.

To stop it we used bricks, but they lasted only a few minutes so we stacked the bricks one after another. When the bricks were gone, we had to turn it off because the building wall would be next.

We called it a KG-1000—"K" for Korad, "G" for gas, and "1000" for the number of watts of power. Compared to our billion-watt ruby laser that may seem insignificant. But the ruby's billion watts occurred in a tiny instant of time whereas the C02 was going on continuously and that's why it burned through everything.

So this KG-1000—now "G" for gorilla—was like King Kong, 30 feet long. The heart of our gorilla was a long glass tube down through the center of which the laser beam flowed. Actually, there were two tubes side by side and connected by beam-bending mirrors at the back end. Out the front came the beam. The two tubes and the turning mirrors made it a folded laser. And because of the glass, it was fragile and subject to breakage.

My job was to install the laser, teach the customer how to use it, and not leave until I was sure they would pay. Fortunately, the customer, General Atomics, was only 100 miles away, down south in San Diego. (Nowadays, General Atomics is famous for their Predator drones that bring terror to terrorists with Hellfire missiles fired at them.) We carefully shipped King Kong in a big air ride truck down the I-5 freeway in the middle of the night to its destination at General Atomics on top of Kearney Mesa just west of the I-5. Trying to keep my time with the King to a minimum, I begged out of setting it up in the customer's lab. Korad workers did the honors. Better the workers than me to get blamed for injuries in moving Kong, I thought.

The General Atomics scientists told me their intended usage was to drill oil well holes. Jerry Shane, Korad's esteemed salesman who sold the laser, had told me their application was materials research. Such vagueness was a tipoff to military secrecy and they were definitely military. I suspected the oil wells to be a cover story.

Anyway, the delight on their faces was apparent when I told them how fearful I was of it and of how nothing stopped it. My oft repeated apprehensiveness seemed to delight them; I was certain they would pay and I got out of there in a laser flash, leaving them with their new toy.--

Nuclear Research

Perhaps the most important research Korad was involved in was its collaboration with Lawrence Livermore Lab in the still-ongoing quest for limitless energy via nuclear fusion. The K1500 and K2600 oscillator amplifiers (lasers) were needed to make "plasma" measurements—but more on that in a minute.

Just imagine the potential for good that such technology promises. Imagine an environment free of the smog and acid rain that result from the burning of fossil fuels. Imagine rain forests and coral reefs recovering from the damage they have sustained in recent years. Potentially unlimited energy could be supplied by thermonuclear fusion—the manipulation of atomic particles in an environment contrived by science involving *truly unbelievable* temperatures. It's basically creating a small sun or star here on Earth to produce heat, and then using that heat to generate electricity. While nuclear fusion is not yet a reality, we were proud to be part of the research.

In my Smart interview, I asked him about that effort in the mid-1960s to 1970s; did he recall the laser process involved, Thompson Scattering? "Sure. What we did was build a laser that produced four pulses, one joule per pulse, each pulse one microsecond apart—and Q-switched to get the power up with pulses lasting nanoseconds. That was pretty good." It was inconceivable for the layman, and still fairly inconceivable to me in retrospect.

Before we return to Smart's specific recollections, let's take a closer look at nuclear fusion and *plasma.* In searching the Internet for a layman-friendly explanation of the process, I settled on a surprisingly straightforward offering by MIT's Plasma Science and Fusion Center:

"The adage 'opposites attract' is true of electrically charged particles as well as people, and is the reason why fusion reactions are so difficult to achieve. Since all atomic nuclei have a positive electrical charge, they tend to repel rather than attract each other.

Consequently, a tremendous amount of energy is required to bring the nuclei close enough for the strong nuclear force to bind them. For this reason, fusion reactions generally occur at temperatures of the order of twenty to one hundred million degrees. At these temperatures, matter exists not as a solid, liquid, or gas, but in a fourth state called 'plasma.' The task of confining even a wisp of superheated plasma fuel has proven to be the greatest challenge in fusion research."

You can see I wasn't exaggerating about those temperatures. Now let's get back to Korad's manager of scientific laser systems, Don Smart, and the process of Thompson Scattering.

"The Korad laser was used for probing, with laser pulses, a plasma to measure its temperature. You had a quartz window and you fired a ruby laser pulse down through the plasma; it emerged out the other side of the plasma. We measured temperature by the frequency shift of the laser as it passed through the plasma."

Pretty fascinating stuff, and all the more interesting when one recalls the jeers and guffaws from the scientific establishment when a relatively unknown West Coast industry scientist named Theodore Maiman first came up with a working laser. Well, OK, they begrudgingly acknowledged, he might have beat us to it, but we did all the preliminary work, so he doesn't deserve credit. Maiman lived long enough to experience ringing vindication on that score.

With the ending of the Vietnam War, military business tanked. So at the end of Union Carbide's ownership in the last months of 1971, Korad laid off about fifty of their military people.[107] Carbide had sold the military division to North American Rockwell (now Boeing).

Next to go were the scientific lasers, as worldwide research on ruby lasers died out. The industry moved on to more efficient and powerful lasers such as YAG, excimers, pico- and femto-second pulses. Gradually over the years, Korad was to concentrate on industrial laser machines like resistor trimmers, welders, drillers, and markers.

---

[107] 50 military people let go according to Jim Scaroni who joined Korad Jan. 2, 1972. This made him wonder about how secure his new job was.

Chapter Fourteen: Who's at the Helm?

S. Donald (Don) Sims

A Short Review of Important Events

Donald Sims, as president of Hadron, a New York-based company of twenty people, purchased Korad from Union Carbide for a paltry $1.2 million in early 1972[108] with a down payment of merely $150,000 borrowed from their New York bank as working capital.[109] They had an estimated five years to satisfy that debt to Carbide and approximately three to reimburse the bank. Assuming prevalent rates at that time of eight to fifteen percent interest, one can

[108] $1.2 million in 1972 was a fraction of the 1968 value of the $3-5 million discussed earlier.

[109] Per an interview with financial executive Paul Ratoff, the first Hadron employee at Korad. He was later promoted to chief financial officer (controller).

estimate monthly combined bank and Carbide payments of about $30,000.

The owners of Korad had been over the years: Union Carbide, then Hadron/Sims, Xonics, and a reconstituted Hadron. But Korad was the brand name for all laser products throughout the company's existence. According to *Laser Focus* magazine,[110] Hadron, with their newly acquired Korad, constituted the largest solid-state laser company in 1973 with annual sales of $4.5 million. Hadron was then laboring under debt payments amounting to about eight percent of its monthly sales. A bit high, but they had not diluted ownership when borrowing money. Another factor in their favor was that they could be highly profitable, making almost all components in-house. *Laser Focus* had to acknowledge, however, that the purchase left Hadron "saddled with heavy debts."[111]

Ed Young, Maiman's controller and long-time Carbide man, told me that when acquiring companies, Carbide never accepted notes or stock; it was cash deals only. So for Carbide to take this deal strongly implied it was a "fire sale." There had just been a change in Carbide's top management, with Birney Mason, the big boss, retiring, and evidently the new management did not want to be bothered by a small and troublesome department like Korad. Young said I was probably correct.

Why did Carbide take the deal in late 1971? They well knew that Maiman and his second-in-command at old Korad, Fred Burns, had each attempted to purchase the company four years earlier. So why didn't Carbide sell it to them?

Pursuing this question in an email to Maiman's widow, Kathleen, I inquired as to why Maiman didn't buy back Korad. I received the following reply.

"At that time, Ted was fighting an acrimonious divorce, made more painful because he cared so deeply for Sheri, his 13-year-old daughter. His wife Shirley wanted to forever be linked financially to Ted, an arrangement that would give her considerable control. From what Ted told me, and his family confirmed, Shirley was off-the-chart impossible! Ted gave up everything he had—all the cash, stocks, real property, even his clarinet, for the divorce he badly wanted."

---

110 "Hadron Expanding," *Laser Focus* (Apr. 1973): 4.

111 "Sims Steps Down," *Laser Focus* (Oct. 79): 40.

As for Dr. Burns' decision not to pursue the purchase of Korad, I can only speculate that he was tied up financially with his venture capital-funded company, Apollo Lasers. Dr. Sims told me that Burns had tried again to buy Korad almost a decade later when the curtain was coming down on Xonics' ownership of Korad.

Until I interviewed Sims, I had harbored an unfavorable view of him because in 1972 he had fired me[112] from my position at Korad, a position I had held for seven exciting and productive years. Subsequent events satisfied me he had done me a favor by letting me go, so at length, all was forgiven and I came to be deeply impressed by Sims' intellect, charm, and honesty.

Born in Brooklyn, he had attended the renowned Erasmus Hall High School, as did a number of luminaries such as Barbra Streisand. Sims went on to earn simultaneous bachelor's and master's degrees from MIT in 1958. He rounded out his qualifications with a doctorate in physics from Purdue in 1963. His doctoral dissertation revolved around the building from scratch of a working ruby laser using only Maiman's sketchy description. One of his primary resources was Charlie Walker's famous photos of Maiman with his new toy, but it must be noted that these photos were intended for news releases, not science. It's interesting that another Korader, Dr. Jim Boyden, constructed a ruby laser in the same time period using Walker's photos. And, as we mentioned earlier in the book, so did Bell Labs in New Jersey. At that time if you wanted a laser, you had to make it yourself, nobody was selling them.

Korad Jumps On, Then Off, the Microelectronics Bandwagon

It was a full decade before the personal computer became reality in the mid-eighties, and it was two decades before the Internet. The big semiconductor companies were Intel, National Semiconductor, Fairchild, Hewlett Packard, Texas Instruments, Motorola, and IBM, the first four of these being headquartered in Silicon Valley[113], the hub of the worldwide semiconductor industry. Naturally, as Silicon Valley was

---

[112] My boss the marketing manager who took over after Bill Thurber's promotion to general manager, Stu Zuck, delivered the bad news to me about my termination. He said Sims claimed engineers could sell. To Sims' credit he quickly changed his mind and hired back sales manager Jim Kessey.

[113] Silicon Valley is south of San Francisco down to and including San Jose.

only an hour's plane ride from Korad, it became one of their most coveted sources of potential clients, with the greater Boston area, called Route 128, a close second.

Hadron had purchased Korad from Carbide in the belief that laser-based, silicon wafer scribers were a potentially huge market. These scribers were used for making semiconductor chips, the heart of microelectronic devices and now a component of nearly everybody's favorite tool or toy. You can visualize a silicon wafer as city blocks laid out in streets and avenues. The blocks are the chips and the streets separate the chips. In the silicon wafer business they call both streets and avenues just streets. Now think of the silicon as window glass[114] that is to be cut into squares. To do this, you use a diamond-tipped pencil to scribe the glass or silicon. Then it is "broken" into square chips to become the brains of your tools and toys.

Unfortunately for Korad and their laser scriber competitors, most notably Quantronix in New York, another way to scribe and break wafers was developed that used diamond saws. These were far cheaper and cleaner. The sawed and cut edge was sharp and clean without any laser debris or burn. Further, the laser's second step, the actual breaking of the scribed wafer, was not required at all. As Phil Schmidt, Korad's vice president of the mid-1970s put it, "Things changed rapidly in the semiconductor industry when it came to what goes on in wafer fabs [i.e., fabrication]. Equipment in wafer fabs could become obsolete in as little as two years." When the dust settled in the mid-1970s, diamond saws had beaten lasers out of that particular market. That development, coupled with the 1974 recession, was a powerful blow to Korad.

The seeds of the company's troubles had sprouted in the early 1960s with the laser's birth when people, most notably Maiman himself, called it *a solution in search of a problem*. Korad simply took on too many problems and failed to specialize. Union Carbide, during that period, favored rolling as many dice as they could, especially where the possibility of major money beckoned. But by now it was a dozen years later and Korad was spread too thin over too many products, offering aging options to too many industries. Years upon years of inadequate investment in product improvement were like an untended garden going to weed. Well-financed competitors were

---

[114] Glass contains silicon and is actually an oxide of silicon.

coming out with better gizmos, and many of them were staffed by Korad people. Fresh money enabled new products to be sold at lower prices than Korad could match because these companies enjoyed the advantage of start-up monies that offset their losses.[115] The price war between Korad and its three spinoffs—Fred Burns' Apollo Lasers, Jim Boyden's Holobeam, and the author's Quantrad[116]—eventually drove all four out of business.

So at this point it was clear to Sims that Korad had too many irons in the fire and it was time to pull a few out. One of the first products to go were the military lasers, a decision given greater impetus by the ending of the Vietnam War and the sale of the military division by Carbide to Rockwell just prior to the sale of Korad to Hadron. Next to go were the scientific products such as the K-1, K-1500, and holography. These products were serving a dying market as research using ruby lasers was no longer of interest to scientists. However, the surviving product line, industrial lasers, was more than enough to support the company.

Earman[117] noted in our interview: "Industrial systems such as resistor trimmers and welders were all fairly simple, profitable, and repetitive non-custom systems. Scribers were a product that, in a growing industry, you could build on a cookie-cutter basis. They were all identical. And there was volume production. So Korad built lots and lots of inventory. You'd go down the halls and they'd have workstations stacked up with components."

When Earman began his job at Korad, he was working overtime on the scribers. Korad came up with a plan to "sell" scribers. They would

---

[115] An outstanding example was Korad's competitor, Holobeam in New Jersey, staffed by Korad engineers led by Dr. Jim Boyden. Year after year they suffered million-dollar losses but had $10 million in their kitty. Koraders joked they would last ten years.

[116] Quantrad got into industrial lasers when the author joined them in 1972 after being terminated by Korad. Floyd Pothoven then came aboard as chief engineer, a double promotion from being a lowly engineer at Korad. I set out to put Korad out of business and I succeeded with lots of help.

[117] Jim Earman's opinions are informed by many years of experience owning and operating laser companies. His first company specialized in laser wafer scribers. His partners were Koraders Phil Schmidt and Jim Scaroni. They sold their company in the 1980s for over a million dollars each. Earman then founded Jimani in Ventura, California, and still operates it as of this writing as a laser marking job shop and equipment supplier.

send them free of charge to semiconductor companies like IBM, Intel, Motorola, HP, Fairchild, and National. Earman said, "We were building as if there were no tomorrow—shipping, installing, and then months later, sending me out to pack them up for return. They believed they'd make a mountain of money on the scribers."

Now let's have a look at a factor external to Korad, the dog-eat-dog world of sales. Hal Walker, whose adventures ranging off the Moon were discussed in an earlier chapter, was an effective sales manager for Korad. In charge of sales in Silicon Valley,[118] he generated orders. But he had lots of problems selling against competitors like Long Island-based Quantronix for scribers and against Boston-based Teradyne for trimmers.

As Walker told me in our fourth interview, "Quantronix and also Teradyne gave us some trouble with their automated systems. In those days we had to compete in a sales buyoff. What happened was the customer wouldn't actually buy our system. They gave us a conditional purchase order for a system, but we'd have to take it to a performance-based qualification buyoff. They typically had a Quantronix demo machine in another lab doing the same thing. Customers were interested in laser lamp life, downtime, and throughput.[119] And even though I think our Korad machines performed on the quality side, they did have problems with staging that caused trouble."

The stage problem to which Walker referred was a poorly designed inaccurate device, the "stage," that moved the silicon wafer under the laser. While Korad insisted on designing and building it in-house, Quantronix purchased their stage from an expert supplier. Korad's was amateurish by comparison and Korad's management refused to improve it—just one more nail driven into the company's coffin.

I asked Walker if being a minority conferred any advantages in his sales job—not in terms of affirmative action, but simply the day-to-day interaction with clients. Then I added with a grin, regarding his African-American heritage, "Because you stand out."

---

[118] "Personnel," *Laser Focus* (Dec. 1973): 59.

[119] All solid-state lasers back then were energized by lamps. "Downtime" was when a machine did not work and "throughput" was how many parts it produced, typically in an hour.

To this he replied also with a smile, "Yeah. I stood out. I was tall.[120] If there was a problem, people tended to offer me leeway, and I could speak to them about the problem and what I could do about it either on the spot or in some way to come back and take care of it. They gave us that grace, such that they didn't say, hey, take this thing and get out of here."

Quantronix, for example, was just selling specs while Walker went out of his way to make friends with his customers. Teradyne[121] people, however, were much more refined; they were Northeasterners, dressed in suits and ties, and very well adapted to business meetings. He said, "They were classy compared to Quantronix."

Walker concluded with, "Also, there was an advantage to my style and my looks. I fit in more with what people were comfortable with—because if a customer had a problem (with their laser scriber or whatever), I could usually fix it right there on the spot."

Scribing Goes Out – Marking Is In

The best and easiest place in the world to sell silicon scribers was in Silicon Valley, south of San Francisco and in and around San Jose. It was bad luck for Korad and their laser competitors that the diamond saw ruined their scriber business. But fortunately, at the same time, along came an incredible opportunity for hungry laser machine manufacturers: lasers for marking. As long as you had the right laser, these models scribed an indelible mark on virtually anything.[122] Initially the marks were identification numbers, letters, or barcodes on high-value products like semiconductors, medical devices, and even nuclear bombs. These folks were willing to pay a lot of money for IDs. Lasers were expensive, so it was a nice marriage of buyers and sellers. Counterfeiters were thwarted, and manufacturers could identify employees responsible for bad parts and take the necessary corrective action.

It is relatively easy to understand how a laser marker works because it can be thought of simply as a supermarket laser scanner

---

[120] Walker was six foot four inches tall, just 70 mm short of two meters.

[121] Teradyne, in the Boston area, was helped considerably in their laser resistor trimming business by Korad's enormously talented Dr. Jim Overbeck, laser genius, and Mogens Raven, software guru. Time frame was late 1970's.

[122] YAG lasers are used on metals and CO2 lasers on plastics.

equipped with a laser powerful enough to vaporize molecules and etch a mark. Laser markers created almost permanent identifications. The machine was normally comprised of three parts: a computer, a laser, and optics[123] to scan and focus the beam. The machine could also feature a small enclosure to house the part being scribed, a little door, and a button one pushed to zap the object.

Scanners were basically invented way back in 1820, but scanning as a process didn't happen until modern times with the proliferation of inexpensive and fast computers. The first use came in 1836 as a way to detect and measure electrical currents. The scanner instrument was and still is called a *galvanometer*, named after Luigi Galvani, who discovered in 1791 that electric current could make a frog's leg jerk. It was an important instrument in the history of a science so dear to our hearts in modern times in the form of electronic games and other more useful devices. Important how? If you want to control something, measure it. This even applies to business: if you want to control people, measure their performance.

Laser scanners to mark alphanumeric characters on parts were recognized early on as a major market. Indeed, the countries of Germany, France, and Canada were willing to finance the necessary research and development as well as subsidize the companies involved in the work. It is for this particular reason that at my company, Florod, I decided not to enter this business; I just did not want to compete against entire countries.[124]

[123] Marker optics featured primarily a flat field lens with the General Scanning XY galvo scanner.

[124] Florod was founded in 1974 by my partner, Floyd Pothoven, and me.

James (Jim) Overbeck, 1978

Wasted Potential

Korad decided to enter the laser-marking business with the help of Dr. Jim Overbeck, formerly an astrophysicist at MIT,[125] where he was known for his work on X-ray stars. He was assisted on the marker project by the talented Korad engineer, Don Smart. Together these two designed and built an effective laser machine. Laser markers really did turn into a big business, responding to a world market clamoring for upwards of 30,000 units per year, according to marketing expert, Gary Sheriff[126] at that time.

Overbeck's baby sported two General Scanning G-300 galvos. He even built the electronic drivers for it, and the company wrote the controlling software in-house. This was very important, as the unit cost was reduced by twenty percent of the sales price, making Korad highly competitive (and profitable)[127] in the marketplace. By the rules that govern business, that twenty percent dropped like a stone to the bottom line, known as profit, thus increasing it by twenty percentage points. So if you were making, say, ten percent profit, you would triple it.

---

[125] Massachusetts Institute of Technology is near downtown Boston and next to Harvard University. Unfortunately, Dr. Overbeck passed away in 2009 of lung cancer, although according to his widow, Anne, he had never smoked.

[126] Private communication. He is located at Shertec Inc. in Southern California and conducts seminars on the entire field of laser scanners.

[127] If profitable, why did Korad have money problems? At least three possibilities: high debt, recession, and especially bad management.

A computer loaded the program from a punched paper tape which made an ungainly clunkity-clunk sound as the tape loaded eight bits per clunk into its memory. The paper tape was about an inch wide and could be hundreds of feet long. Holes were punched in it, with each hole or lack of a hole representing a digital "bit," a zero or a one. These days we load data millions of times faster, but this was, relatively speaking, the Jurassic Period. Still, we couldn't know what incredible breakthroughs were to come, so we struggled along with what was then the cutting edge of technology.

Overbeck designed an extra large lens[128] to focus the beam so that it would more effectively vaporize and thus mark. Smart called it a "really effective laser marker," but went on to say that Korad could not produce them because of a lack of money. Printed circuit boards were the parts they needed, but these, of course, were more expensive, so the company decided to cobble together their own version called "wire wrap boards." Wires on the circuit board were "wrapped" using an electric motor tool similar to an electric drill that spun the wires over little posts sticking up out of the circuit. Unfortunately, the wires wrapped around the posts would come loose with age and then the device would fail. Then the customer's tech would take the covers off to fix the problem, see those primitive wire wraps, and there went Korad's reputation. As the main customers for laser markers were the semiconductor houses, they knew that wire wrapping was for prototypes *only;* it was common knowledge in electronics companies that the wires could easily come unwrapped.

Here is what Jim Earman had to say about the situation: "Korad was so desperate for money that they sold Overbeck's prototype. I think it went to Motorola in Florida. Then they built a few more after that. Everyone in the company realized we were really onto something with laser marking, but the management failed to improve it. They had Overbeck building more wire-wrapped board type units. Therefore, they were a nightmare to keep running. We could see the writing on the wall; what with pay cuts and shipping junk to get cash, it was a hand-to-mouth existence."

---

[128] An F-theta flat field lens is the heart of all two-dimensional galvos (2D). And a variable focus beam expanding telescope imbedded in the laser optics is the essence of all 3D galvos.

So the vice president of sales, Phil Schmidt, requested a meeting with President Sims. He said to him, "Don, we're not trying to threaten you, but Scaroni, Earman, and I are all ready to bail out of here. Our feeling is that the only thing that's going to save this company is the laser marker. If you're not going to do anything with that, we're gone." 1Sims agreed and promised to dump money into the laser marker. But it didn't happen. Eventually, Sims was to make it happen, but for the three mutineers it was too late; they had departed to form their own company, Laser Identification Systems (LIS), not surprisingly, to make laser markers.

Good People Led by Challenged Managers

The nasty hi-tech recession of 1974 weakened Korad. This was in spite of the efforts of seventy hardworking, talented employees, including Phil Schmidt, head of marketing; Hal Walker and Jim Scaroni in sales; Dr. Walter Koechner,[129] head of engineering; Jim Overbeck, Don Smart, and Mike Weiner also in engineering; and Jim Earman, head of customer service. These were all highly capable, masters in their field, as their subsequent professional successes amply illustrate. Lack of talented technical people was certainly never a problem at Korad—in the entire history of the company, no less! Management was another matter.

At the twelve-year mark, Korad had built up quite a large base of installed lasers, resulting in a spares and accessories business of around $1 million per year. Jim Stimpson, who was in charge of this business unit, was to make his life's work the selling of laser accessories. I was astonished at how much money he made working for others doing this. Another factor in Korad's favor for survival—but not growth—was an extremely broad product line accommodating both scientific and industrial markets.

---

[129] Walter Koechner, *Solid-State Laser Engineering*, (New York: Springer-Verlag, 1976). Now in its sixth edition, this book, along with the dispersal of Koraders throughout the entire solid-state laser industry, was Korad's legacy to the laser business.

Walter Koechner, 1969

He Who Spread Korad Technology Far and Wide

Dr. Walter Koechner is well known and respected in the laser industry for the six editions of his book, *Solid-State Laser Engineering*. Koechner composed his manuscript while employed as engineering manager. His book contains all of Korad's recipes for cooking up highly reliable lasers. Koechner was the inventor and designer of most of Korad's (YAG) solid lasers, but of course not the ruby solid lasers. His book was to become the "bible" of laser designers throughout the world. It spread Korad technology far and wide becoming the gold standard of solid laser markers. All laser companies, even the gas laser ones like Spectra Physics and Coherent, copied Korad's solid laser designs[130] from a combination of expatriates and Dr. Koechner's laser recipe book. This was the K-Y2, see the ad on page 132.

Acle Hicks, Korad engineer/physicist and subsequently father of Coherent's Avia laser, a powerful solid state device, had this to say: "Koechner's book in my opinion is the best practical book on laser engineering that exists. I used it a lot and I know other engineers did also. As to the design of Korad's YAG laser, when I was there Bud Erickson was the designer and developer. He left Korad in 1967 with the Boyden group for competitor, Holobeam in New Jersey. The design that Erickson came up with at Holobeam was very successful

---

[130]The K-Y2 was Korad's most significant laser and was copied by the world's laser companies. It is described as a continuously pumped, repetitively acousto-optically Q-switched Nd:YAG, and still is in use.

and was taken (not very ethically or legally) to Control Laser by Lee Benson. Holobeam sued Control who then offered Holobeam $25,000 for the rights to use the design. Holobeam foolishly took the money because they did not feel threatened by Control but Control was able to compete on price and hurt Holobeam a lot. Then when Lee Laser in Orlando, Florida, copied the design with enough changes to call it a new design, Holobeam was devastated...they lost most of their YAG business."

Brief Business Backgrounder

At the beginning of 1974, Korad was reported by *Laser Focus* magazine to be grossing $5 million[131] a year. In this same issue the company was said to be the leading materials-working laser outfit, with Maiman's ex-employer, Hughes Aircraft, a probable second. I remember admiring the Hughes cloth-cutting machines for rapidly fabricating expensive custom men's suits, a very good use for a CO2 laser and quite a departure from the military. No doubt they were motivated by the end of the Vietnam War to diversify into commercial.

At the end of the following year, *Laser Focus* reported Korad's system sales at a mere $2.6 million for 1974.[132] Add to that the typical $1 million in spares, accessories, and repairs for a total of about $3.6 million. Furthermore, the article reported Korad as one of the leaders in the solid-state laser systems business, so presumably they were doing as well as the marketplace would allow. Nevertheless, sales had fallen off by about 25 percent in just one year—quite a drop, and probably due to the recession of that year, which *Laser Focus* noted affected the laser community for thirteen months, bottoming out in February of 1975.[133]

Markers were to become a major laser application, a very big "app" that all well-informed people in the business were aware of. Another really big opportunity emerged in 1974: laser barcode scanners initially employed at supermarkets for each and every register. It was a ball game fit for big-time players like Spectra-Physics, who had secured a $9.8 million order from National Cash Register for upwards of 4,000

---

131 "Industry and Construction," *Laser Focus* (Jan. 74): 36.

132 "Year end report," *Laser Focus* (Dec. 1975): 48.

133 "Review and Outlook," *Laser Focus* (Jan. 1976): 10.

laser scanners.[134] It was a tough game for a little-leaguer like Korad to compete in, so they stayed out of supermarket barcode scanners.

Management Gets Creative: Out-of-the-Box Thinking?

In 1976, when an order for a laser machine would come in, an empty box would be shipped to the customer. Armed with the shipping documents for the empty carton and using these documents as security (collateral), the company borrowed money. This type of lending is called "factoring" in the business world and those who do it are factors. So that's how Korad generated cash flow to run their business.[135] Of course, the factor's fees are steep. It is said that cash is not king; it's the *emperor*! The empty-box ploy was highly risky because if the factor were to discover it, he could terminate the relationship, possibly putting the company in bankruptcy.

In our interview Don Smart said, "I didn't see Korad lasting long. Don Sims and Paul Ratoff were running the company down the tubes, as far as I was concerned. They were doing these strange things trying to get money into the place by conning the factor. I knew they were doing this because I was at the Detroit arsenal installing a laser. The people at the arsenal told me, 'That was very nice, Don, thank you very much. What should we do with the other laser in the next room?'"

Smart was shown a big box displaying a sign that read, *To be opened by Korad personnel only*. So he phoned Paul Ratoff, who told him not to worry about it and to have them send the box to the next customer so Korad wouldn't have to send another box.

Dumb and Dumber

Smart's story was not unfamiliar to me, as I had heard exactly the same in my interviews with Korad's other key employees: Jim Earman, Jim Scaroni, Jim Stimpson, and Phil Schmidt. Smart went on to relate how Ratoff and cohorts took a shipping company executive out to dinner one night, got him well and truly oiled with a bunch of drinks, and then talked him out of a stack of shipping documents so they could expedite their shipping process.

---

[134] "Spectra to Begin Deliveries," *Laser Focus* (Oct. 1974): 14.

[135] In those days almost all business was done on credit. Payments by Korad's customers were made directly to the factor, not to Korad. Cash up front was considered an insult.

A Deal You Can't Refuse ...

In yet another "creative" business practice, Korad retained as an outside contractor, and not as an employee, a former Air Force colonel, one Jack Moore, a karate black belt. His duty was to negotiate with the company's suppliers who were owed money for equipment or parts they had supplied Korad. His goal was to pay only ten cents on each dollar owed. If he was successful in getting this ninety percent reduction, he was rewarded with a ten percent commission. Moore was known at Korad for having arm wrestled a supplier for money. Moore won. Smart overheard Moore on the phone doing his routine and passing himself off as Korad's president. The colonel wanted a membership at the Beverly Hills Country Club, so the company provided this expensive perk in his hometown.

The ten percent wheeling and dealing was costly as the suppliers predictably raised their prices to cover their losses. They also began to demand payment up front. Others simply refused to supply Korad with any more of their products. This forced the company to either redesign their products or find another supplier. But their poor credit rating would tend to lead the new supplier to similarly demand cash up front. The effect of all this was to drive Korad even deeper into the clutches of their factor. It was like an anaconda slowly tightening about its twitching victim, choking off its final breaths.

Korad's reputation did not escape notice in the pages of *Laser Focus* magazine. "Last year (1976) Hadron closed many of its debts with creditors who had agreed to settlements as low as ten percent."[136]

Almost every organization, whether public or private, occasionally has financial trouble. As a laser company owner and president for thirty-six years, I have certainly had my share of bad times with no available lenders. I never reduced pay, but I was forced to delay some paydays. However, I did factor (accounts receivable). Most factors squeeze you with a yearly contract, and escape from the anaconda requires a lot of cash, which is hard to get in strapped circumstances.

Don Smart tells an amusing story of how one day there were two guys taking the company's copying machine out the back door. He bolted for Ratoff's office, who was able to placate them with a check, thus getting the copy machine returned. Shortage of money was

---

[136] "Korad Losses Said To Total $3million," *Laser Focus* (May 1977): 30.

obvious to all the employees. But when their pay got docked by half, that's when the exodus began. Bill Rundle was one of the first to leave, joining Dr. Mal Stitch at Hanford Nuclear Labs in Washington state. Also Hal Walker and Mike Wiener departed for Hughes Aircraft with some of Korad's best.

However, many top employees stayed, including Jim Earman, Jim Scaroni, Phil Schmidt, and Don Smart. They all related the same bleak story of retroactive pay cuts and sob stories from management. Korad allowed events to occur which had only one outcome: a company lacking its best employees will be in deep trouble. Most company managers would have laid off the marginal employees. But at Korad there were few if any marginal employees to lay off.

Jim Earman held out until 1977, while his friends, Scaroni and Schmidt, remained another year. These three alone were sufficient to run a company, as a few years later they started their own very successful laser company and emerged as millionaires with enough to start other ventures. Korad had promised that everyone who stayed would be paid, and they lived up to their promise. Working in Korad's favor was an old rule of thumb in business that states twenty percent of employees do eighty percent of the work.[137]

Hal Walker and the Korad Mafia at Hughes

After working ten years at Korad, Walker got a job offer in 1974 from Hughes Aircraft's military group that he simply could not refuse. He joined Hughes as a supervisor in 1974, ironically joining the company where Maiman had invented his ruby laser. When Maiman left Hughes to join Quantatron, a total of eight made the jump; now Hughes got back even more.

Walker began work in El Segundo down the street from the Los Angeles airport, LAX, where Hughes had built a one-million-square-foot facility primarily for the production of laser weapons. The Cold War with the Russians was heating up, and the Vietnam War had proven to the American generals and admirals the outstanding effectiveness of smart laser weapons. Hughes had a plethora of competent scientists and engineers, but a dearth of experienced laser production people.

Walker was courted by Hughes, who sought him for his production and management experience, and who hoped that he would lure some

[137] The 80:20 rule, a.k.a. Pareto's Principle, see Wikipedia.

of Korad's laser production people to staff their imposing El Segundo facility. The Koraders became known at Hughes as the *Korad Mafia*. Each member prominently displayed in their office a 1968 group picture of the Koraders in front of their Santa Monica building. It became a badge of honor, envy and respect.

The Korad Mafia at Hughes

With the help of the so-called Korad Mafia, Hughes was able to manufacture laser weapons by the tens of thousands, becoming the world's largest laser manufacturer. It is highly likely that this gargantuan stockpile of smart laser weapons played a part in winning the Cold War and thus averting nuclear war.

"Mafia" Photo – 1968[138]

#	Korader	Hughes Department	Year Rejoined Hughes
1	Dr. Rick Pastor*	Chief Scientist Head of Chem Physics Dept.	1967
2	Dr. Tony Pastor	Chem Physics Dept	1967
3	Dr. Hiroshi Kimura	Chem Physics Dept	1967
4	Ken Arita	Chem Physics Dept	1967
5	Jim Linn	Engr Development	1972
6	Dr. Bill Buchman	Engr Development	1972
7	Bernie Soffer	Engr Development	1974
8	Hal Walker	Production Test Supervisor	1974

[138] An enlarged copy is available at www.maimanbook.com

9	Jim Myers*	Production	1974
10	Jim Earman	Production	1977
11	Rick Towns	Production Test Supervisor	1984
12	Mike Weiner	Product Development	1977
13	Ed Gregor	Engr Development	1979

*Deceased

By 1983 Walker had become a third-level manager at Hughes with the title of manager of laser development programs. He was then two steps up the management ladder from supervisor. His duties included designing development models and *proof of concept*, also known as *proof of principle*.

Hadron Has Had It

Various problems and a business opportunity led to the acquisition in early 1977[139] of Hadron/Korad by Xonics, Inc., located in Van Nuys, a suburb of Los Angeles. Xonics was controlled by an ambitious promoter named Bernard (Bernie) Katz. His company was in the business of making medical equipment, most notably a machine that detected breast cancer via mammography. They did $24.5 million in annual sales. Korad's production capability looked attractive for making their equipment. Xonics also had an excimer laser prototype and an X-ray machine that Korad could work on. Sims, remaining as president of Korad after Katz took over his company, described Katz as a somewhat shady operator, a financier, and "the kind of guy who would do what he had to do basically to win." Sims went on to say he was a small man, fast-talking, a super-promoter about fifteen years older than him. "He was very smart, very quick, and had no moral compass at all."

Laser Focus reported that Xonics purchased 84 percent of Hadron/Korad for $841,000 in stock, not cash. Sims retained his twelve percent stock ownership. The nearly million-dollar debt to Union Carbide which paid for the 1972 sale of Korad to Hadron had been paid off by H.L. Federman and Co., a New York venture capital firm, in exchange for Hadron stock.[140]

---

139 "Xonics' Acquisition of Hadron," *Laser Focus* (May 1977): 28.

140 *Laser Focus* (Dec. 1976): 4.

Korad Meets Its Nemesis

Another article in the May 1977 issue of *Laser Focus* stated that Sims was to report to Dr. Earl W. Brian Jr., president of Xonics.[141] So, the big boss of Korad was now Dr. Brian. I will discuss him at this point as he was a colorful character and the subject of many news articles, none of them particularly complimentary.

Earl W. Brian Jr.

Born in 1942, Dr. Brian was a combat surgeon in the Vietnam War in a medical unit attached to the Central Intelligence Agency. He performed his duties so well in the agency's Phoenix Program that he was awarded the Silver Star, Bronze Star, and more. Upon his Army discharge with the rank of major in 1970, he did postgraduate work at Stanford University's medical school where he treated one of Governor Ronald Reagan's campaign managers, Ned Hutchinson. Brian so impressed Reagan that by late 1970 he became Reagan's California Secretary of Health and Welfare Services. He also became friends with Edwin Meese, a friend of and attorney for Ronald Reagan. Meese was to become the attorney general of the United States during the Reagan administration.

A few years later, Brian's father died and left him enough money to enable him to become a venture capitalist, a buyer of companies in which he pumped up stock values to make a "killing." One of these companies was Bernie Katz's Xonics that later was to come under Dr.

[141] "Xonics Acquisition," *Laser Focus* (May 1977): 28.

Brian's control. Both Katz and Brian were to become stock "pumper uppers," reminiscent of the robber barons of the free-wheeling nineteenth century.

According to Sims, "Earl Brian was politically connected. He was very smart and completely detached from any ethical basis at all—that just wasn't part of his world. It's not that he did not know right from wrong; it's that he didn't care. He was different from Bernie Katz who didn't really know right from wrong; Earl Brian *knew*."

The promotional efforts of Brian and Katz brought in orders until 1977, when the U.S. government came down on Katz, as Brian had bailed out some six months before. The Securities and Exchange Commission, the U.S. government's powerful watchdog over U.S. publicly owned businesses, slapped Xonics with a permanent restraining order[142] for violations of the antifraud securities laws. Xonics had been issuing press releases trying to boost stock prices with exaggerated or bloated information.

I recall reading a series of exposé articles in the *Los Angeles Times* back then that Katz drew a salary of $20,000 a month and carried his wife on the payroll for an equal amount. So that's a half million dollars every year that he was milking Xonics. This was over ten times what a typical company president took home. The *LA Times* was making a point that he was bleeding his company dry. But they were mainly attacking him for inflating stock value, thus cheating his stockholders.

Here's an amusing insight into the infamous Mr. Katz. The following is from an article that appeared in *Forbes* magazine, January 21, 1991, written by Julie Pitta, entitled *The Show Goes On*. "Bernie Katz, the P.T. Barnum of high-tech stocks, is back. In the 1970s he hawked Xonics, Inc. and its supposedly improved X-ray machine. In the 1980s he pushed Helionetics, Inc. and its blue-green laser intended for the (Ronald Reagan) 'Star Wars' defense shield (FORBES, Apr. 22, 1985)."

The Katz Empire crumbled, with both companies going bankrupt. (A Google search would likely provide considerable specifics should a reader be interested.) A common theme for Katz and Brian in their business world was to be the power behind the throne. Brian was the chairman of the board but CEO only once, at the beginning of his business career at Xonics. For the CEO's job he soon chose Dominic

---

142 Copy available on the Internet. Dated April 7, 1977.

Laiti (pronounced "Lie-tee"), who became Sims' boss. Sims described Laiti as follows:

"Dominic Laiti was a smart guy, a pretty straight arrow in that sense. Not necessarily a nice guy, but sort of straight—not dishonest. It's curious to find yourself in a business situation with these guys and suddenly realize what it is that you're dealing with; it's a very frightening kind of thing if you're a straight arrow. I'm sort of a straight arrow kind of guy. So I'm uncomfortable with things of the nature they got involved in."

In my interviews with the two Korad presidents, Sims and his successor, Bob Schlesinger, it struck me that both resigned their jobs when Brian and Laiti put pressure on them to falsify company financial records, a crime for which many company presidents go to jail. I daresay every publicly owned company president is more than well aware of this threat to their freedom. As a president of a private company, I knew of it and rejoiced that I was reasonably safe from a federal prison simply by owning my company. Further, I didn't have to put up with pesky stockholders demanding dividends.

In 1979-80 Dr. Sims, as president of Korad and under the direction of Brian, picked the name Compulaser. Korad remained as a division of Compulaser and as a brand name. The other division was Digicon in Mentor, Ohio, a computer company. Sims became president of both. And Brian took Sims' established name, Hadron, to be his own; he renamed his company Hadron, a nifty way to get rid of the bad reputation of Xonics.

By March 1980, Sims had gotten fed up with Brian. As Sims told me in our interview, "It just wasn't working out. It was a lot of pressure, and I walked away from the thing." Korad's new owners needed money because they were, according to Sims, under a lot of financial stress. The owners of Korad's Santa Monica building wanted to sell their property, so they paid Hadron to relocate across the Santa Monica Mountains to Canoga Park in the San Fernando Valley. For Sims the long, arduous twenty-two-mile drive from his home in rush hour traffic on the jammed 405 freeway over the Santa Monica Mountains was an important factor in his subsequent departure from his job. Sims resigned in 1980. He had been president of Korad for eight years, longer than Maiman, who had served five years, 1962-67.

Here is how my interview with Dr. Sims ended: "I'll tell you one more story, the last one for today. At the very end of things, after I had walked away from the company but was still sitting on the board of directors, they scheduled a board meeting. I just ignored it. This guy Steve Bolen, Brian's financial man, showed up at my house. He asked my wife, 'Where's Don? We need him for the board meeting.'"

In reply she said, "He's sick. He can't see anybody." Sims didn't want to be part of what he perceived as illegal, so he resigned. He said, "The good part was I was able to sell my stock in a hurry." And he was no longer an *insider.*

As an insider (company president and member of the board of directors) he could sell his stock only in spaced-out, small amounts due to SEC laws. But after resigning, his stockbroker could "roll and roll" the stock. Unfortunately, the stock price was going down fast, not because Sims was selling, but because "things were screwed up." As he went on to say, "I didn't want any part of it. I didn't want any inside information." He got out in the nick of time with a tidy sum of money and no possibility of being thrown in jail.

The long reign of Dr. Don Sims had come to an end as head of Korad. He left after having succeeded in focusing on just a few promising products, including markers, welders, and drillers. The company's Carbide debt problems had ended when Xonics purchased the company, but debt payments were just replaced with corporate overhead that had to be paid to the fat cats at the top, now known as Hadron.

As a company president, the technically competent Dr. Sims' main problem was producing consistent profits. Quoting a 1977 *Laser Focus* article about the Xonics takeover of Hadron and the end of Sims' control of it,[143] "Hadron has had a troubled history for more than a decade, with accumulated operating losses estimated at $3 million despite sales that were said to have peaked at $3 million two years ago: sales last year barely exceeded $2 million."[144]

In answer to the question posed at the beginning of this chapter—"Who's at the helm?"—the answer would clearly have to be Dr. Earl W. Brian, with Dominic Laiti to execute his orders.

Next, in our last chapter we tell the story of how the dubious duo of Brian and Laiti managed to destroy Korad. The historic scandals of the Reagan presidency make for informative, if not amusing, reading.

---

143 "Losses Said to Total $3 million," *Laser Focus* (May 1977): 30.

144 $2 million in 1977 would have nearly $8 million worth of buying power in 2013.

Chapter Fifteen: The End Game

Robert J. (Bob) Schlesinger and his wife, Sylvia

The Next-to-Last President of Hadron's Compulaser Division

In early 1981 Bob Schlesinger took the reins as Compulaser/Korad's president. Dr. Don Sims had quit the job the year before[145] and Dominic Laiti then took over,[146] but it was time consuming, time that could be more productively spent in Virginia rather than California. Laiti had been through two presidents. The first was Charan Lohara, who quit so that he could return to his native India, and the other was Lowell Benjamin, who had previously been with a Beverly Hills maker of scientific instruments.[147] Unfortunately, history does not record if

145 "Sims leaves Compulaser," *Laser Focus* (Nov. 1979): 110.

146 "Sims steps down as Hadron president," *Laser Focus* (Oct. 1979): 40.

these two gentlemen quit because of Brian and Laiti's illegal business practices. And I do not count them as presidents because their times in office were so short.

At the time of my interview with Schlesinger, he was suffering from Alzheimer's disease, so his story was mostly related by his wife, Sylvia, a highly energetic and professional journalist of many years' experience, who met her future husband while writing a story for one of her client publications on the solar power industry of Orange County. He owned an electronics company making devices to control electrical power generation from sunlight, both for residential and for utilities companies. However, power did not usually come directly from the now-ubiquitous solar panels,[148] but from thermal power, where focused sunlight heats a liquid that in turn generates electricity.

At any rate, their paths crossed at that point and they married, after which Schlesinger promptly sold his seventy-person business, Rho Sigma,[149] to a New York Stock Exchange company, Watsco, for a mere \$150,000.[150] The solar industry had lost U.S. government support and funding, rendering his operation less attractive on the market. He was quite lucky to get out when he did as the industry almost disappeared a few years later under the Reagan administration, which believed that meddling with private enterprise was ill advised.

Perhaps Schlesinger's sense of humor helped him find a buyer in Watsco. His story was told in an October 15, 1977 *Forbes* magazine article, where in reply to an interviewer's question as to whether or not he would sell his company, Schlesinger replied, "Only dummies go around slamming doors in peoples' faces." Becoming bored with boating, he immediately began job hunting, complaining every day to Sylvia, who knew better than anyone that her husband would never be happy on the sidelines. Schlesinger's interest was ignited one morning over coffee when he spotted a Hadron help-wanted ad in *The Wall Street Journal*: *Wanted: Company president to preside over two operations, one in Canoga Park, California, and the other in Mentor, Ohio.* The opportunity appealed to him as both companies sold

[147] "Sims leaves Compulaser," *Laser Focus* (Dec. 1980): 110.

[148] Known as photovoltaic (PV), discovered by Albert Einstein in 1905 for which he was awarded the Nobel Prize.

[149] Rho Sigma were his initials, RS, in the Greek alphabet.

[150] \$150,000 in 1980 is the equivalent of \$420,000 in 2013.

electronics—always a plus in his world—but beyond that, one company even produced high-tech lasers, an irresistible opportunity for an electronics engineer. Each company employed about twenty but was struggling, and this posed a real challenge.

Schlesinger regarded his prospective new boss, Dominic Laiti, as a suave and articulate businessman who was sure to go places. In his capacity as head of these two plants, he was to report to Laiti, who was based in Virginia with Hadron, a company primarily selling software. From Laiti's perspective as Hadron's president, Schlesinger was a highly talented, successful high-tech company founder and president who would look good to investors.

After accepting his new position, Schlesinger quickly found the company to be truly in trouble. There were no orders coming in, and although he had been a successful president of his own company in the past, he had no experience saving failing companies. Quoting Sylvia, however, "Bob was enough of an ego to think he was truly hired to turn these companies around. He liked a challenge and welcomed the idea of playing Mr. Fixit. It was when Brian and Laiti were obviously ignoring everything he proposed and showing little interest in anything other than doing an IPO [initial public offering] to go on a public stock exchange that he became very uncomfortable and suspicious."

A Few Good Men

At least there was money available to hire a few people, including John Derzy, an extraordinary salesman, and David Collins, a hotshot engineer. Hadron also provided funds to build up three laser marker demo units so that Derzy could leverage them to get orders. These provided demonstrations for customers, and often involved the temporary loan of a demo system for placement in a good prospect's business. Compulaser/Korad sported a staff of engineers consisting of electrical, mechanical, laser, and a chief engineer. In addition they had managers of production, purchasing, sales, and field service.

John Derzy

Derzy had been an electronics test technician at Rho Sigma prior to selling Schlesinger on promoting him to sales. While working at Rho Sigma, he attended night school in Hollywood to learn broadcasting at the Don Martin Broadcasting Institute, but subsequently (at Korad) he fell in love with lasers and decided to abandon broadcasting.

Salesman Derzy said he saw Dominic Laiti rather frequently when the latter visited Compulaser/Korad in the Los Angeles suburb of Canoga Park. He described Laiti as tall and arrogant with a distinguished thatch of graying hair, a man who always wore a fine suit and tie and came across as a polished veteran of senior management. Derzy commented that Laiti had a tough side but was never a screamer or a fist-pounder.

Schlesinger's responsibilities at Compulaser were a nightmare. He had hired engineer Collins to solve problems that kept cropping up with products that had been shipped to customers only to promptly fail. One such incident resulted in being sued by a customer in Indianapolis who had purchased a laser marker to ID bearings they manufactured. It would operate for about fifteen minutes and then shut itself down for an hour or two before starting again. Collins made the long trip to Indianapolis to meet with the plaintiff's attorneys. He guessed the cause was simply overheating. A water chiller was being used to cool the laser, and Collins quickly discovered an air trap in the chiller that prevented water from flowing.

He excused himself and slipped off to a local hardware store to purchase a simple plumbing fitting for $25, returned, installed it, and attempted to start the laser. It seemed timid at first, burping and spitting a bit of water out of the top, but then it ran flawlessly. The attorney looked at him and asked, "Now just why didn't they do this sooner?" Because so many of the Korad lasers were exhibiting difficulties, Collins suspected sabotage, which I'll discuss again later in this chapter.

Characters at Hadron Headquarters

The Schlesingers recalled how Earl Brian would come into board meetings in Virginia from 1981 to 1982 smoking a huge cigar. He would take off his shoes, sit down at the conference table, and prop up his feet. He was somewhat of a grotesque character whose nicknames bore that out: "Cash" and "Big Daddy." Schlesinger detested Brian but had to put up with him only at the East Coast meetings where Brian was the kingpin in his position as chairman of the board.

Schlesinger recalled discussing with Sylvia events that occurred at one of the headquarters meetings that Dr. Edward Teller attended. Teller is best known as the father of the hydrogen bomb and a key player in President Reagan's *Star Wars* initiative. The plan was to build a defense against Russian ballistic missiles that involved heating them to destruction using extremely high power American lasers.

Dr. Teller also became known for being the model for *Dr. Strangelove* in Peter Sellers' movie of the same name where he played the role of a mad scientist chief advisor the President of the Unites States. If it weren't about nuclear war, it would be as outrageously funny as his Inspector Clouseau or Pink Panther movies

Teller's contribution to Star Wars arguably helped America end the Cold War with the Soviet Union. If so, the money spent on the laser was a very good investment indeed. Having the father of the H-bomb working on Star Wars very likely got the attention of the Soviets. The vast inventory of America's "smart" laser weapons supplied by Hughes and employed so effectively in Vietnam likely gave them additional pause. It was a wise decision on the part of Russia to throw in the towel and end the Cold War. If you can't beat 'em, join 'em!

Peter Sellers (as Dr. Strangelove)

Dr. Edward Teller

Garbage

An important topic of discussion at the Hadron board meetings was taking their companies public to enable big profits. They looked for companies to acquire, inflated the stock price, and then they cashed out by selling to investors. Sylvia Schlesinger called the companies they bought *garbage shells*. They didn't put much money into them, or to put it in the language of business, capital investment was low. They were merely going to pump them up. But she said back in the 1980s there was a lot of this going on, which brings to mind Xonics/Korad's ex-prime mover, Bernie Katz. It's interesting that he controlled a famous Star Wars company, Helionetics in San Diego, and that Dr. Teller was associated with it. But like Katz's other company, Xonics, it went belly up as well.

The Infamous October Surprise Scandal

Unbeknownst to Schlesinger, Dr. Earl Brian, the prime mover of Hadron, was said to have been deeply involved in the 1980 U.S. presidential election when Ronald Reagan was running for president against the incumbent Jimmy Carter. As a personal friend of Edwin Meese,[151] who worked for Ronald Reagan, Brian had allegedly been recruited to carry $40 million cash to Paris for payment to Iranian

---

[151] Edwin Meese became the seventy-fifth attorney general of the United States (1985–1988).

terrorists.[152] The angry Iranians had imprisoned fifty-two American Embassy workers in Tehran since November 1979.

Now that the November 1980 presidential election was nearing, the Republicans allegedly became nervous that President Jimmy Carter was going to get the embassy prisoners released and win the election. Reagan was holding Carter responsible for this fiasco, and it was such a hot political issue that Reagan was certain to become president provided the hostages were not released until after the election.

The Republicans did not want to risk release of the hostages just before the November elections—thus Earl Brian's trip to Paris to deliver the bribe. His mission was at the heart of the *1980 October Surprise*. The opponents have long denied the October Surprise; however, the Iranian president during the hostage crisis, Abolhassan BaniSadr, acknowledged these reports years later. Whether strictly accurate or not, it is a historical fact that Reagan's two terms in office were plagued by scandals, making President Harding's Teapot Dome look like an afternoon tea party by comparison. But Reagan and his successor, George H. W. Bush (the Sr.), won the Cold War, preventing millions of deaths in a hot nuclear war, similar to the one that nearly resulted from the Cuban Missile Crisis of the John Kennedy presidency.

The 1980 October Surprise was, and still is, a highly polarizing political issue. I include it here because, as will be explained, it led to the downfall of Korad. The scandal's truth will likely remain forever unresolved. The U.S. House of Representatives investigated it twice and in 1993 found no truth to it. I find it highly interesting that the chairman of the House investigation, Lee Hamilton of Indiana, a Democrat no less, could find no conclusive evidence, thus indicating there was none. Hamilton's committee read tens of thousands of documents over a ten-month period, conducted over 200 interviews, and spent over a million dollars.

More Sabotage

Getting the job done was the whole ball game, and alas, all the (brand name) Korad products had across-the-board problems with laser machines refusing to function. Welders were impossible to keep running reliably at the customers' places of business.

---

[152] Numerous references are on the Internet, including Wikipedia.

Superman engineer Dave Collins was called upon by his boss, Schlesinger, to rescue the company by addressing these issues. Collins was a single man who often spent virtually all his spare time, including weekends, dusting off and reading voluminous old documents from the file cabinets. Sitting on the floor in the file room, he had the old-time lab notebooks spread around him like a desert island in the middle of an ocean.

By carefully studying the old docs, he became curious as to the nature of the specified optics and began to suspect that the welder blueprints had been altered by persons unknown, resulting in the lasers' subsequent failures. The sabotage involved only the laser optics, nothing else. The deed had been done by an optics engineer, otherwise known as a laser engineer. The perpetrator had no mechanical knowledge or else he would have altered the drawings of parts. The same goes for electronics. In my opinion the criminal was obviously a disgruntled engineer, probably one who had been severely reprimanded with a dire warning and was about to be fired. Collins corrected the prints and the problems vanished.

In our interview, Collins said that Laiti did not care about this at all. Because Laiti was unconcerned, neither was Collins; the culprit escaped.

Two unusual and interesting laser projects that people worked on in its final days were a welder for Idaho Nuclear and a driller for the National Bureau of Standards.

Laser Welder for Nuclear Reactors

Idaho Nuclear Test Facility ordered a Korad laser welder to fuse stainless steel thermocouples to nuclear fuel rods so that they could keep track of temperature. If a reactor were overheating, they could shut it down before meltdown. In the middle of the mountains of Idaho, west of Idaho Falls, the Navy tested their nuclear reactors to the very point of destruction, purposely overheating them, which made it an exciting place to visit. David Collins recalled being instructed at the very beginning of his visit, "If you hear this little siren go off, run like hell!"

Last of the Big Ruby Lasers

The second project involved the National Bureau of Standards in Colorado, which purchased a Korad driller. It was no ordinary driller;

indeed it was a highly unusual one, probably never again duplicated as it functioned using ruby laser technology, soon to become obsolete.[153] It could drill holes through quarter-inch stainless steel, over six times deeper than modern commercially available solid lasers. The project involved zapping holes in a fifty-gallon drum that was used to perforate postage stamps for the U.S. Post Office. Hundreds of holes were required in a single drum. A laser driller had to be used because mechanical drill bits wore out. Ruining just one hole in the drum made the whole drum worthless because one error in a sheet of stamps caused a ruinous tear in every sheet made by that drum. So that drum would have to be thrown away. Laser drilling, on the other hand, presented no such obstacle; you could be sure most drums were good.

Schlesinger's Out

Mastermind Earl Brian's plan was to take his companies public on a stock exchange and thereby make a killing, whereas Schlesinger's intention was to turn them into decent, functional companies. It was not to be. No sooner did Schlesinger arrive at Compulaser but he was informed that they were taking the company public.

To build up the company in the eyes of the public, Brian and Laiti wanted him to cook the books by falsifying the company's financial statements, mainly the inventory value. Schlesinger told them it was useless inventory and he would not approve it. So he quit after having worked there less than a year. It was early 1982 when he and Sylvia went to live in England where he got his doctorate (he called it his "union card") and began work as a college professor. Hadron had soured him on the business world, and he never again returned to it. But he didn't hate the business world; he even taught business. Sylvia said in our interview of April 10, 2010: "He admired well-run businesses and entrepreneurship. He just had an aversion to crooks."

During the author's interview with Dr. Schlesinger, his Alzheimer's was not so advanced that he couldn't come up with this zinger: "You understand, you can't take a company public saying that it's got assets and resources that it doesn't have. It would be defrauding the investors, a federal crime which can result in prison." He went on to

[153] In laser language it was an oscillator with two amplifier heads to produce 400-joule pulses in conventional non Q-switched mode.

say, "I wouldn't sign and I quit. Understand that."[154] Sylvia was totally shocked that he came up with this.

Collins Steps In

When Schlesinger quit, he recommended to Laiti that Collins be promoted to fill his position as president, and Laiti did exactly that. Collins inquired as to whether a pay raise went with the promotion and was told no, but he wanted the position so he accepted. After Schlesinger left, the following events happened after Collins took control.

Partnering with IBM

Sig Kupka, born in Germany of an American mother and a German father, was thoroughly Americanized. Employed by IBM in Germany, he had been successfully promoting IBM robots there. Transferred to the States, he came to know John Derzy, and they got the idea of combining their lasers and robots. Derzy said Kupka was the first to introduce IBM robots to the United States, but of equal importance was the first use of an IBM PC computer in an industrial machine, the Korad laser. Both Derzy and Kupka introduced the combination IBM and Korad laser marker at the 1982 midyear International Manufacturing Technology Show, IMTS, in Chicago. During the 1980s it was commonly said that the "P" in "PC" meant "personal" and that meant it was unsuitable for use in a machine. How wrong that idea turned out to be! Now the PC is the computer of choice to run all sorts of things.

For their IMTS tradeshow booth in Chicago, Derzy and Kupka hired two Chicago Bears football cheerleaders. The girls' job was hosting a laser-marking demonstration. The laser was to etch names of those visiting the booth onto handout key chains. These women were to men like flowers to bees. Besides being professional cheerleaders for a major football team, they were also models. Throw in a laser and the brand-new IBM PC computer, and you had powerful male attractants. Besides, the key chains were of excellent quality and personalized with each particular guy's name. A pretty girl at the head of the line typed the name, and the girl at the end handed the goodie to him with her "gorgeous hands," as Derzy called them. She went to the extra trouble to wipe off the laser residue. This was the first semi-automated robotic

---

[154] Both Sylvia and I were shocked at this most lucid statement.

work performed by an IBM PC. There was, however, trouble caused by the mob of people blocking nearby exhibits. Angry complaints were directed at the show management, so the line was split in two and separated by an aisle between.

The outstanding success at IMTS led to a deepening of the relationship between IBM and Compulaser/Korad. A flood of sales leads began to flow into Derzy's eager hands—so many, in fact, that close to all their incoming business stemmed from IBM. Sales leads would not come from the Internet for another twenty years. In those days they were acquired from costly advertising, tradeshows, and magazine articles usually helped by an expensive ad. But then, as now, the very best leads come from word-of-mouth recommendations. If someone you trust says something is a good buy, it gets your attention. Leads from IBM were good as gold. Derzy admired Collins for his dogged pursuit of IBM. Derzy said without his support and pushing, the involvement with IBM would not have happened.

Competition and the Marketplace

In 1982, the year of high hopes at the company, there were no more than five hundred laser markers in use worldwide. But there was an expectation of selling that many in that year alone. Competition was limited to only three main players: Control Laser in Florida, Quantrad in nearby Torrance, and, of course, Korad. There weren't many differences among the three. Control and the Korad machine were look-alikes; both were packaged in identical enclosures, Hoffman NEMA boxes. They even had the same color, blue, with white insides. Quantrad was quite different; there was no enclosure, just a laser on a rail that dangled over the part to be marked. The neat thing about this clunky design was you could just stick whatever it was you wanted to mark under it and *voila*! You had it! But Quantrad had issues with laser safety because the laser produced peak power pulses up in the tens of kilowatts—enough to blind a person. So users had to wear laser safety glasses. And there was the legal liability of bystanders suing for laser eye damage, real or imagined. Performance was the same for all three, quite simply determined by the number of alphanumeric characters you could write in a second. Because differences were minor and prices essentially the same, it came down to who was most adept at locating customers.

Compulaser/Korad was fortunate in having the inside track with IBM. Despite the exorbitant price of laser markers, there was a ready market for them in jet-engine turbine-blade marking of IDs, bearings in the turbines, and lots of other expensive items. About these things buyers would often say, "Permanent mark sounds good, laser mark looks good, I'll buy it." Even though it was far and away the most expensive technology on the market, they could somehow rationalize buying.

As lasers were expensive, the things they were used on tended to be high priced also. Derzy offered the example of circuit boards used in the electronics revolution of the twentieth century. They were easily copied and sold as counterfeit. Cheap knock-off parts used by the military angered the Pentagon and Congress, causing manufacturers considerable pain. So they rushed to use laser markers. And commercial electronic chip companies did the same. According to Derzy, the market price for a laser marker in 1982 was $83,000. In 2013 that's worth about $200,000. Having an excellent reputation, the Korad management decided to sell theirs at $50 more. Why fifty dollars?

Korad had been suffering from brutal price wars with their main spin offs, Apollo Lasers and Holobeam, for nearly a decade. Now in the mid 1970s they perceived they had a superior product, the marker, and the lowest cost. Remember the marker, a souped-up store barcode scanner, is quite simple, with only a few components. The most expensive were the laser and the galvo with its associated software and computer. (A galvo is simply a limited-rotation motor that twitches a mirror on the end of the motor's shaft for laser-beam steering or scanning.) Programming the galvo was an exceedingly difficult job. Compulaser/Korad had the best software engineer I have ever encountered (I hired him after they collapsed). His name was Mogens Raven. What made him totally unique was an ability to accurately give the date of completion of when he would have a software program written.

Anyway, he wrote the Korad marker software. This allowed the company to buy the guts of a galvo, the mirrors and the electronics driver, for about $3,000. The price of the galvo with software to the laser integrators (those who buy laser components and assemble them) was nearly $20,000. If you didn't make the laser, your cost was another $20,000-30,000. Add the two and you get $40,000-50,000.

Add another $20,000 for the DEC PDP-11 computer and you are too close to the $80,000 market price for the whole marker shebang. Therefore, there was insufficient profit for those who had to buy both laser and galvo. So they stayed out. And those who could make either laser or galvo made a so-so profit (but not enough for me when I owned Florod Lasers in the mid 1970s).

So Korad was in the driver's seat. They had by far the lowest marker cost, around $20,000, and their competitors knew that. The $50 higher price was less than one hundredth of one percent of the $80,000 market price, which served to put the competitors on notice that they were not going to wage another price war.

Given this rosy picture of the Korad market position, why did they fail? Short answer: Earl Brian was making tons of money off his stolen Promis spy software and was to dump Compulaser/Korad so as to get it out of his path to riches. More on Promis follows.

Collins Calls It Quits

As president, Collins had been well aware that Brian's Hadron wanted to dump it. He had gone to South America and landed a million-dollar order for laser markers, but Hadron ordered him to reject it, so he quit and went to work for Disneyland's Imagineering, becoming an expert in amusement park rides and accidents. Korad was failing because Hadron was making far more money selling software than they were with the Korad hardware. They could duplicate programs for next to nothing and sell them for hundreds of thousands of dollars. It made no sense to waste their precious time on lasers.

The Inslaw Scandal and Going Out with a Bang!

In the late 1970s and early 1980s Brian and Laiti's Hadron began selling to the U.S. Department of Justice a software program called Promis. In harsh language they stole it[155] from a small company called Inslaw, for whom the scandal is named. In more gentle language they merely appropriated it. The Promis program was highly useful and valuable to the spy agencies of the world, such as the CIA, Israel's Mossad, and the Royal Canadian Mounted Police's spy agency. It allowed them to keep track of people involved in their cases, but it

---

[155] Maggie Mahar, "Beneath Contempt, Did the Justice Dept. Deliberately Bankrupt INSLAW?" *Barron's National Business and Financial Weekly* (Mar. 21, 1988).

wasn't just spy outfits; federal prosecutors in the Department of Justice found it indispensable, and terrorists did too—possibly even bin Laden's Al Qaeda.[156]

In spite of three trials in federal courts and a Congressional hearing, no convictions ever resulted. Crime sometimes pays, but we cannot call this a crime because of the fact that they got away with it. But they also got away with alleged murder.[157] One Danny Casolaro, a journalist, was about to release a story linking Inslaw, the Iran-Contra Affair, and the October Surprise.[158] The night before he was to conduct his final interview, he was found dead in his hotel room. His briefcase and incriminating papers had disappeared. In spite of being twice ruled a suicide, he had a dozen or more slashes on his body. Furthermore, his friends reported he had told them if he were found dead, it would be murder and not suicide.[159]

The argument supporting Hadron's merely "borrowing" Promis was that the program had been developed using mostly government money, therefore it was in the public domain. But Inslaw had created an improved 32-bit enhanced version that was clearly Inslaw's property under the Copyright Act of 1976. It was this version that Hadron sold. Inslaw found out about the alleged theft when the Canadians came to them asking for a French-language version. Inslaw knew the Royal Mounted Police were not *their* customer.

Sylvia Schlesinger posed the question, "Why did the United States Justice Department help Earl Brian criminally take over the little Inslaw company?" Answering her own question, she said, "It was Brian's payback for helping Reagan get elected with his nefarious work in the October Surprise." Inslaw was the payoff.

On the phone, Dominic Laiti told Inslaw's president, Bob Hamilton, that he wanted to acquire Hamilton's company. Hamilton replied that he didn't want to sell. According to the lawsuit filed by Elliot Richardson, Laiti's reply was, "We have ways of making you sell." If this is true, it would establish Mr. Laiti as Dr. Brian's attack dog.

---

[156] Carl Cameron, "Excerpt of Fox News Special Report" *Fox News* (Oct. 16, 2001).

[157] Michael C. Ruppert, "Promis," *From The Wilderness*, Special Edition (Sept. 2000).

[158] Elliot L. Richardson, "A High-Tech Watergate," *The New York Times* (Oct. 21, 1991).

[159] I can relate to this. What with my sister's suicide, I'll never do it either!

Inslaw's attorney, Elliot L. Richardson, was an exceedingly high-power lawyer. As President Nixon's attorney general, he had resigned rather than obey Nixon's order to fire prosecutor Archibald Cox, who was in hot pursuit of Nixon about the Watergate scandal. This incredible guy, Richardson, was suing the Department of Justice over which he had been its big boss as attorney general. But he lost, both in the courts and in Congressional hearings.

As I promised at the beginning of my sordid Inslaw Scandal Story, my reason for relating it here is that the purloined software program, Promis, became for Hadron an extremely lucrative[160] product beginning in the early 1980s. As will now be explained, Hadron was to totally abandon Compulaser/Korad. Because actions are said to speak louder than words, what I hear them say is they basically had a far bigger and fatter fish to fry than mess with such a minnow. It seems that Hadron was in the business of selling the Promis software, and the company became wealthy under the names Analex and, after 2007, as QinetiQ. In the early 1990s, when Hadron changed their name to Analex, they were a $100 million-a-year company.

An Ignominious Ending

Within six months after the IMTS show in the summer of 1982, Derzy was thinking to himself, wow, this Korad is finally getting traction! It's being resurrected by IBM. It actually got some good, solid customers, and sales are being made. But Hadron was not interested in Derzy's sales or sales potential. Seeing the handwriting on the wall, Derzy began to look for another job. He called an IBM friend, Jerry Sculls, a senior buyer in the purchasing department, to tell him what was going on in the company. Derzy told Sculls he absolutely was not moving to Virginia.

Sculls then suggested he approach Korad's competitor, Quantrad's president, Yoram Peleg. Furthermore, Sculls told Derzy to tell Peleg, "If you hire me, I'll bring you a lot of IBM business." However, Derzy didn't feel comfortable with this "boasting" strong-arm tactic, so his Quantrad interview did not result in a job offer. When Sculls learned of this, he called Peleg and said, "If you hire Derzy, I'm bringing you some business. What are you going to do?" Derzy got the job. He had

---

[160] Software can be highly profitable because manufacturing cost is close to zero but design costs can be expensive and are generally unpredictable.

brought in the big gun. As a salesman, Derzy knew very well that people do business with those they know and trust—in short, friends. Therefore, good salesmen make friends with their customers. Korad was a good training ground for Derzy's life in the business world and a fine launching pad into his lifelong career in lasers.

In early 1983 Hadron was shutting down Compulaser/Korad and moving the company to Virginia, the so-called Beltway of Washington DC. But the move was without people; none of the Koraders were willing to relocate away from sunny Southern California. Derzy said the vans came at night and out it all went. Stanley Cherubin, purchasing manager in charge of materials, was the last man standing. He shut out the lights after the move and was the last on the payroll. After the company's stuff was trucked away to Virginia, Hadron tried to hire people but failed. Then they put the physical assets up for sale with drawings and inventory but failed in that as well. Cherubin decided to buy the Korad "junk" as it was a really good deal at pennies on the dollar, and so it all had to be shipped back to California. When Cherubin died in the mid-1990s, his son Max continued to sell Korad spare parts to the time of this writing in 2013.[161] Derzy estimated the Korad parts inventory was purchased for three cents for every dollar it was worth. He claims that Stanley and son resold the spare parts for seventy-five cents on the dollar—a nice markup!

May 22, 1983, marked the death of Korad/Compulaser because that's the date that the SEC delisted (abandoned) the stock.[162]

Dr. Earl Goes to Prison

Brian pulled his last shenanigans with the United Press International, a very big and well known company, when he acquired it as a bankrupt company and then illegally bumped up the stock. He had gone too far and they finally got him. In 1996 he was convicted on ten counts of fraud and sentenced to four years in federal prison, where he served the entire four years. His conviction came after twelve years of Republican Presidents Ronald Reagan and Bush the Senior. When the Democrat, Bill Clinton, rose to power in January 1990, Brian found out

---

[161] Max Cherubin at Panatron in Rancho Cucamonga, California.

[162] "Issue 83-64", US Securities and Exchange Commission (sec news digest), (April 4, 1983).

that even a highly decorated U.S. Army combat surgeon and decorated Vietnam War hero was not above the law.

The reason he got away with all of this as long as he did was he obviously had friends in high places, mainly the CIA, the Department of Justice, and, most notably, Edwin Meese, President Ronald Reagan's attorney general.[163]

A Final Wrap-Up by Phil Schmidt, Laser Entrepreneur

The source of the following is from my interview of the late Phil Schmidt of June 2010.

"The lasting accomplishments of Korad came from all of the glorious pioneering efforts made by so many Korad people during the 1960s and into the early seventies. They had a long-lasting impact on industry after industry as technology development spawned from Korad's work on lasers. These talented people labored to further technology, out of which has grown so many technologies that we see today. That's the legacy of Korad."

Korad's robust and simplistic designs spread throughout the industry. The inventiveness of the Koraders contributed to the success of many companies. After all, the "death" of companies is just the end of a brand name; the people go on to work under other names and hopefully do a better job with what they learned.

In Conclusion

Although Korad came to an ignominious end, it served to seed the laser industry with highly trained engineers and scientists as well as production and sales professionals. Among the engineers, I'd place Dr. Walter Koechner absolutely at the top for technology transfer because of his seminal cookbook on how to make solid-state lasers, *Solid-State Laser Engineering*, published by Springer-Verlag and now in its sixth edition. Second is Dr. Jim Boyden with his inventions of the HP laser and ink-jet printers. Next would be Len DeBenedictis and Acle Hicks for technology transfer to Coherent in Santa Clara, California. Hicks was the father of the highly successful Coherent Avia laser and DeBenedictis was their longest-serving president (eight years). Last would be Lee Benson and Jim Hadwin, who carried away Korad's cw Q-

---

[163] James Ridgeway, "Software to Die For," *Villiage Voice - Moving Target*, an interview of Elliott Richardson (Sept. 24, 1991).

switched YAG technology first to Holobeam in New Jersey and then to Control Laser in Florida. From there it spread to all the solid state laser companies.

For production I'd place Hal Walker at the top for his efforts at Hughes as a third-level manager in helping them become the number one producer of military lasers, contributing to the end of the Cold War and being instrumental in quick military victories in the two Gulf Wars.

For salesmen there are Jerry Shane, Phil Schmidt, and perhaps me. Jerry, with the help of his four outstanding sons, created a 150-person electronics subcontracting company, Qual-Pro, in Carson, California, with over $10 million in revenues. Phil, with his two Jims—Earman and Scaroni—had the vision to start a laser company, Laser Identification Systems, that inscribed ID marks on silicon semiconductor wafers, allowing significant improvements in productivity. Each of them came out millionaires when they sold to Lumonics in Canada. The laser company I cofounded with Floyd Pothoven, Florod, produced the world's first laser eraser to correct errors in the making of semiconductor photomasks, chips, and liquid crystal displays (LCDs). Sixty of these in the hands of Japan, Inc. set them up in the early 1980s to dominate the LCD TV business, because back then nobody could make LCDs without killer mistakes. And the worldwide semiconductor industry saved hundreds of millions, possibly billions, of dollars by significantly speeding up their research and development and time to market. The electronics revolution was accelerated by our efforts. Even though Korad lasted only 20 years, it set a standard of excellence in laser design and reliability that survives to this day.

INDEX

Appendix I: Simplified Laser Background Information

How a Ruby Laser Works

Boasting a capacitor bank that stored energy fed to it by a power supply and a flashlamp similar to those found in a flash camera, the ruby laser's capacitor would store a charge of sufficient electricity, at which point the flashlamp would be "triggered" to fire, causing the capacitor bank's stored energy to dump into the lamp. The resulting blinding flash would be absorbed by the chromium atoms in the ruby, which would then be raised into an "excited state". As the atoms return to their normal "ground state", light is generated at a specific wavelength (0.6943 microns). If the ruby rod is located inside of a "resonator", this light is reflected from and oscillates between the two mirrors of the resonator, causing stimulated emission by the remaining chromium atoms, resulting in the emission of a laser pulse through the mirror of lower reflectivity. Chromium added to sapphire makes ruby, and it is the chromium that does the lasing.

Q-Switched Lasers

What made the Q-switch innovation so remarkable was a brilliant modification conceived of and implemented by a fellow physicist and colleague of Maiman's at Hughes Malibu, Dr. Bob Hellwarth. The following elaboration on his innovation is found in Maiman's *The Laser Odyssey*:[164]

"Bob Hellwarth's concept was to temporarily block the buildup of the laser oscillations[165] by isolating the back mirror with an electro-optical switch in the laser cavity until all of the energy is stored in the excited fluorescent level of the ruby, at the upper laser level. The block is then very rapidly removed electronically therefore, a switch. The

[164] Odyssey, 160

[165] Blocking is what a rollup window shade does to sunlight. Oscillations refer to the laser beam bouncing back and forth between the front and back end mirrors common to all lasers (and getting amplified each time it passes through the lasing medium, whether it be CO2 in a gas laser or neodymium in a solid state YAG laser).

gain amplification in the ruby crystal is so high that the oscillation builds up at an extremely rapid rate and, very briefly, rises to an enormous intensity. The result is a laser output pulse with huge (pulsed) peak power that lasts for a tiny period of time.

As an example, a non Q-switched ruby laser may emit a peak pulse power of say 10 kilowatts that lasts for two thousandths of a second in its conventional operating mode. But, in the giant pulse Q-switched mode, the pulse would only last just 10 billionths of a second with a (pulsed) peak power of 100 megawatts 100,000,000 instead of 10 kilowatts 10,000![166]"

Three Kinds of Q-switches

They are known as mechanical, passive and active. The mechanical ones consisted of a rotating mirror, passive Q-switches contained a bleachable dye, and did nothing except await the next flash of the flashlamp, and active Q-switches detected the flash and then the switch opened electronically. What the active and passive Q-switches did much better than the mechanical was raise the laser's power from piddling thousands of watts to a staggering one hundred million watts. Of course, it was just a few billionths of a second flash of (laser) light, but let's not quibble here; that's a terrific amount of power. Therefore, it made Korad lasers extremely desirable to researchers, and they sold like proverbial hotcakes to university and commercial laboratories worldwide. Competitors tended to offer mechanical whereas Korad did passive and active.

The Mechanical Q-switch

It was the simplest, consisting of just rotating or twitching the rear 100 percent reflective mirror in the laser cavity. Every time this moving mirror lined up with the front one, the beam bounced between the

---

[166] It is useful to remember that power in watts is always energy in joules divided by time in seconds. It doesn't matter whether the energy is optical, electrical, mechanical, or nuclear. So energy and power are NOT the same. In Maiman's example there is one joule in the laser pulse which is a little less than a quarter of a calorie of energy. The immense power of pulsed lasers is possible if the burst is terrifically short. In this example, dividing one joule by a hundred millionth of a second yields a hundred million watts. If you got this in your eye, you would be blinded for life. Isn't it amazing what a quarter of a calorie can do? I once smashed into tiny pieces a two-carat diamond gemstone with one joule of laser energy.

mirrors and you got a momentary giant, high–power, Q-switched laser pulse. It is noteworthy that if the mirrors are *not* parallel, no lasing occurs. This means that when the moving rear mirror didn't' line up with the front mirror, no beam bouncing occurred and thus no lasing.

The Passive Q-switch

Another type was the passive[167] dye Q-switch, commercialized at Korad by Bernie Soffer, a brilliant laser physicist and scientist who made numerous discoveries in lasers. Soffer's passive laser switching device was just a transparent chamber, called a cell, filled with a liquid dye that bleached out to become transparent to the laser beam. Bombardment of the dye by photons of light coming from the laser head caused it to bleach. Well within a second the dye restored itself to blocking the beam from bouncing between the end mirrors and was ready for the next cycle of pulses. This laser consisted of the dual end mirrors typical to all lasers.[168] Between them were the laser head and the passive Q-switch. Then there was a power supply to provide electricity to the laser head and a water cooler to remove waste heat. Oh yes, and a slab of metal, called a rail, on which to mount the laser components. Korad sold a lot of these, known as a K-1QP, because the passive Q-switch was inexpensive and produced a powerful[169] 100 million watt, 10 nanosecond duration laser pulse.

Controllable Active Q-switches

The third type of Q-switch provided by Korad was called "active" because electricity was supplied to it to make it work. The big advantage of this Q-switch was control. When you sent the enabling electrical pulse to it, you would get your laser pulse. This electrical pulse caused the Q-switch to produce an electric field which rotated the plane of polarization of the highly polarized laser beam so that it would pass through crossed polarizers. Not familiar with polarization? Think of polarized sunglasses that block glare. That's the idea behind an electro-optic Q-switch.

---

[167] Passive means no electricity is supplied to it, unlike the active Q-switches that require electricity to operate.

[168] The technical name for the end mirrors is Fabry-Perot etalon.

[169] 100 million watts

What does "Q" in Q-switch mean?

What's Q? It's simply energy stored divided by energy dissipated (lost). So a high Q laser has a lot of stored energy and not much lost. The light energy is stored in the lasing material, typically a laser rod within the laser head. Then when the Q-switch opens (goes transparent), zap! You get an intensely brilliant laser pulse. In a billion watt laser that's enough to break down air like lightening does in a thunderstorm. And it produces a crack sound too, like lightening.

Korad designed and made all three Q-switches: mechanical, passive, and active. In the modern world specialty companies make these, and yet others deposit coatings on optics. Korad did it all. By 1964 Korad was making K-1 ruby lasers five at a time on a mini-production line.

APPENDIX II: Peter Mendoza's Story

Mendoza's story, entirely written by him, did not fit in the book but is so interesting and uplifting that I decided to include it here for your reading enjoyment.

Poor Hispanic Background

In 1940, the year I was born, my family was living in Wilmington, California, a waterfront town near San Pedro. Unfortunately, my mother and father did not stay together, and when I was three, my mother, older sister Lydia and I moved northward to Fresno, where my mother found employment as a migrant farm-worker.

As a kid, I enjoyed swimming in the irrigation ditches of Fresno while my mother worked in the fields. We lived in a single-room house, with an outdoor wood-burning stove, shielded on both sides with cardboard, with no indoor plumbing. I have nothing but good memories of those days. At that time I didn't realize we were poor.

When I was eight we returned to Wilmington. There my mother worked in the fish canneries as she continued her struggle to raise

Lydia and me. She somehow managed to send me to Catholic schools, from Holy Family Grammar School through St. Anthony's High School.

I was a "C" and "D" student throughout high school. My mother, with a third-grade education, did not know how important homework was, and unfortunately I took full advantage of the situation. Still, in my family I was among the first generation to graduate from high school.

Upon graduating from high school, I got a job at a fertilizer plant, driving a forklift. While working there, I was contacted by a salesman who convinced me that if I completed his correspondence course in electronics, I would be able to get a better-paying job as an electronic technician. So I signed up right then.

While working on my home-study course with DeVry Technical Institute, I got a job as a mailman for the U. S. Post Office. My mother was terribly proud of me and my mailman uniform. By her standards, I had reached a high level of success.

At the start of the home-study course, I knew nothing about electricity. To my surprise, I was eager to learn about electrons, neutrons, transformers, capacitors, *etc.* I was even motivated to attend college—something no one in my family had ever done. I enrolled at Harbor Junior College in Wilmington, seeking an "AA" degree in electronics. Again to my surprise, I was an "A" student in the electronics and math classes. I realized then that my high-school grades were not a reflection of my intellect, but rather the result of poor study habits and lack of motivation.

I then decided to pursue an engineering degree, and I choose an emphasis in physics. Enrolled at California State College Long Beach, I worked my way through college working as a technician on the Minuteman Guidance System, as a part-time TV repairman, and at Korad as a part-time technician. Upon graduating, I was hired as an engineer at Korad. That was the beginning of my education in laser and optics technology.

Korad – Education

My experience at Korad was both an exciting entry into a high-tech industry and a continuation of my education. When I was in college, Cal State Long Beach had no formal curriculum on the study of lasers. Fortunately the scientists, engineers, and the technicians at Korad became my teachers. Korad was the lucky break I needed to enter the new and growing laser community.

One of the highlights came while I was working for Dr. Bill Buchman on a frequency-doubled YAG laser. Dr. Buchman's design consisted of a temperature-controlled, frequency-doubling crystal placed inside a laser cavity. After a few weeks in the lab and with help from co-workers from other labs, we successfully doubled Infrared (IR) to green light. Dr. Buchman was so proud of his success that we took the frequency-doubled laser to all the other labs to demonstrate IR being converted to green light.

Hollywood was also impressed. Universal Studios purchased a frequency-doubled laser for the film *The Andromeda Strain*, released by Universal and directed by Robert Wise. Dr. Dennis Rice took the prototype laser and turned it into a finished product.

I had a chance to see a preview of the film with some other members of the Korad staff. There was one scene in which the laser appeared to burn one of the actors on his hand. This scene caused Korad's owner, Union Carbide, to become concerned that the company would become subjected to negative PR. Fortunately, no negative publicity resulted.

During the time of transition of Union Carbide to Hadron ownership, I was promoted to manager of manufacturing engineering, a position that was extremely challenging for me, especially given the level of education and experience I had at the time However, even though I may not have been the most qualified, the promotion served as a tremendous educational opportunity. Eventually Dr. Rice left Korad to join Northrop Research and Technology where he offered me a position. I accepted his offer.

While at Northrop I co-authored a few technical papers, and was awarded a U.S. registered patent as co-inventor of a fast, high-energy, electrical laser switch. Also, while I was working for Dr. Rice, fellow Korad graduate Rick Towns and I were given the assignment of serving as consultants to Northrop Electro-Optical Division. The Electro-Optical Division had a serious problem on a multimillion dollar contract with the Air Force for a Laser Target Designator. During the many months of investigation, the Hughes YAG laser used in the Laser Target Designator had become an issue. The interesting aspect for me about this situation was that I represented Northrop in meetings between the two companies, and Hal Walker, another Korad graduate, represented Hughes. Eventually Rick and I were given complete control of the project, as well as credit for rescuing the contract and saving Northrop millions of dollars.

In retrospect, I feel extremely honored that a kid with "C" and "D" grades in high school and only a four-year degree in physics would be made a member of the laser-research technical staff at Northrop Research and Technology Center. I attribute much of this wonderful opportunity to the excellent laser and optical education I received at Korad and to the gifted people who worked there.

Appendix III: Korad's Entrepreneurs

(Koraders who started businesses – a partial list)

1. Walter Koechner, Fibertek, Herndon, VA, lasers and systems for the military and industry. The largest of the Korader's businesses. Over $100M in sales, employing 170 scientists and engineers. Founded 1983 (to present).
2. Gerald Shane, Qual-Pro, the second largest business of them all, located in Carson CA, a subcontract manufacturer. $25M in sales and employing 150, mostly workers. Founded 1971 (to present).
3. Ted Maiman, Korad (1961), Maiman Associates (1968), IDAK (1970), and Laser Video Corp. (1971).
4. Don Sims, Hadron Inc., Westbury, NY.
5. Fred Burns, Apollo Lasers (1968), Playa del Rey, CA.
6. Art Lubin, Image Optics (1968), Santa Monica CA, and Laser Institute of America, LIA in Florida.
7. Phil Schmidt, Jim Scaroni and Jim Earman, Laser Identification Systems, Oxnard, CA.
8. Bob Schlessinger, Rho Sigma, Santa Monica, CA.
9. Hal Walker, AMAN, a philanthropic enterprise in Inglewood, CA.
10. Rod Waters and Floyd Pothoven (1974-1996), Florod Lasers, laser machine supplier, Gardena CA.
11. Floyd Pothoven (1997), International Technology Works, Bellflower CA.
12. Rod Waters (1996-2011), Laserod, laser job shop, Torrance, CA.
13. Jim Earman, Jimani, a laser marking job shop in Oxnard, CA.
14. Phil Schmidt, a manufacturer's representative business in San Diego, CA.
15. Marty Phillips, Non-Linear Devices, a manufacturer's representative firm in Oakland, NJ.
16. Marv Sachse, safety consultant, Marina del Rey, CA.
17. Herb Stein, Chinese laundry and a machine shop, Los Angeles
18. Dennis Mulvaney, Radio Shack franchise
19. Jim Hadwin, MacDonald's franchise, Montana

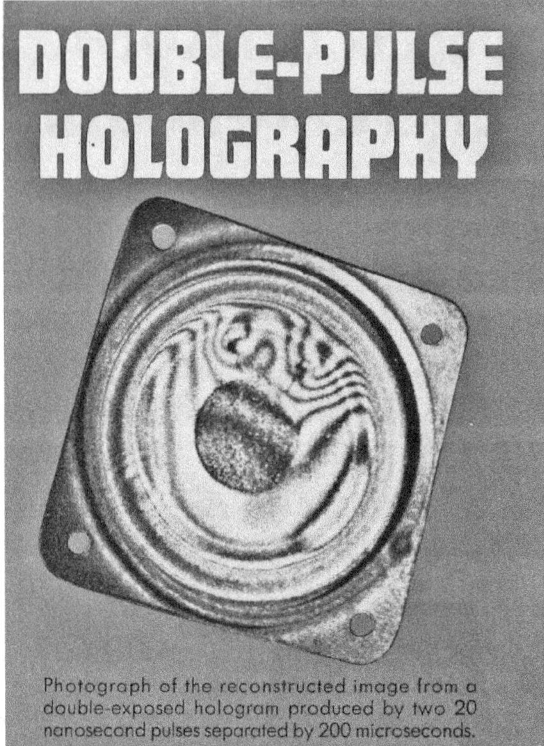

Photograph of the reconstructed image from a double-exposed hologram produced by two 20 nanosecond pulses separated by 200 microseconds.

KORAD K-1QDH

This hologram of a speaker driven at 200 Hz was produced by the KORAD K-1QDH. The holographic ruby laser system uses a special double-pulsed Pockels cell Q-switch. Separation of the giant pulses can be varied over a broad range. Fringes show all of the points on the speaker cone which have been equally displaced between exposures.

Applications for this laser system include non-destructive testing, stress and vibration analysis, wind tunnel aerodynamics, and plasma diagnostics. It is also possible to analyze the spatial distribution, size and velocity of small particles using this high energy system.

For more information call (213) 393-6737, Ext. 224, or write KORAD's Holographic Systems Group today to discuss your applications.

Union Carbide is a total supplier to the industrial community through its Components Department, Crystal Products Department, Instrument Department, KORAD Department, and Semiconductor Department.

ELECTRONICS DIVISION
KORAD® DEPARTMENT

Appendix IV: References

1. "Laser Action in Ruby", **T. H. Maiman**, British Communications and Electronics, Vol. 7, P. 674, 1960. See also Nature, Vol. 187, p. 493, 1960.
2. "Stimulated Optical Emission in Fluorescent Solids II. Spectroscopy and Stimulated Emission in Ruby", **T. H. Maiman**, The Physical Review, Vol. 123, No. 4, p. 1151-1157, Aug 15, 1961.
4. "Biomedical Lasers Evolve Toward Clinical Applications", **T. H. Maiman,** Hospital Management, April, 1966.
5. "Continuously Tunable, Narrow-band Organic Dye Lasers", **B. H. Soffer and B. B. McFarland**, Applied Physics Letters, Vol. 10, No. 10, 15 May 1967.
6. "A Ruby Laser Modified for Pulse-Transmission-Mode Cavity Dumping", **W. J. Rundle**, J. Applied Physics, Vol. 39, No. 11, p. 5338-5339, Oct 1968.
7. "The Ruby Laser: Its Present and Future", **W. J. Rundle and M. L. Stitch**, Laser Focus Magazine, March, 1969.
8. "Design and Characteristics of a Narrow-Pulse PTM Ruby Oscillator-Amplifier Laser", **W. J. Rundle**, IEEE J. Quantum Electronics, Vol. QE-5, No. 6, p.36, May 1969.
9. "Resistor Trimming and Other Micromachining with a YAG Laser", **R. L. Waters and M. J. Weiner**, presented at the Pennsylvania State University Engineering Seminar on New Industrial Technology, July 7-9, 1969.
10. "YAG Challenges Carbon Dioxide in High C-W Power", **Walter Koechner**, Laser Fous Magazine, Sept, 1969.
11. "Performance of a 2GW Ruby Laser Designed for Lunar Ranging", **W. J. Rundle**, EIA-US Dept. of Commerce, Paris Laser Colloquium, Nov 18, 1969.
12. "Analytical Model of a C-W YAG Laser", **Walter Koechner**, Laser Focus Magazine April 1970.
13. "Mode Selection in a Pulse Transmission Mode Ruby System for Holography", **E. Gregor, B R. Guscott, and J. J. Myers**, Applied Optics, Vol. 9, p.1723, July 1970.
14. "Laser Rangefinding", **M.L. Stitch**, Chap F7 in The Laser Handbook, Vol. 2, F. T. Arecchi and E. O. Schulz-Dubois, eds., North-Holland Publishing Co., Amsterdam, 1972.
15. "Present State of the Art of Ruby Laser Holocameras", **W. J. Rundle and T. V, Higgins,** presented at the Los Angeles SMPTE Technical Conference, Sept, 1975.

www.ingramcontent.com/pod-product-compliance
Lightning Source LLC
Chambersburg PA
CBHW051453170526
45166CB00001B/230

9781492842309